Understanding Basic Statistics

FIFTH EDITION

Charles Henry Brase
Regis University

Corrinne Pellillo Brase
Arapahoe Community College

Prepared by

John Stokes
Edutorial

BROOKS/COLE
CENGAGE Learning

Australia • Brazil • Japan • Korea • Mexico • Singapore • Spain • United Kingdom • United States

For product information and technology assistance, contact us at
**Cengage Learning Customer & Sales Support,
1-800-354-9706**

For permission to use material from this text or product, submit all requests online at **www.cengage.com/permissions**
Further permissions questions can be emailed to
permissionrequest@cengage.com

ISBN-13: 978-0-547-18892-8
ISBN-10: 0-547-18892-7

Brooks/Cole
20 Davis Drive
Belmont, CA 94002
USA

Cengage Learning is a leading provider of customized learning solutions with office locations around the globe, including Singapore, the United Kingdom, Australia, Mexico, Brazil, and Japan. Locate your local office at **www.cengage.com/global**

Cengage Learning products are represented in Canada by Nelson Education, Ltd.

To learn more about Brooks/Cole, visit
www.cengage.com/brookscole

Purchase any of our products at your local college store or at our preferred online store
www.cengagebrain.com

Printed in the United States of America
3 4 5 6 7 14 13 12 11 10

Table of Contents

PREFACE.. v

UNDERSTANDING THE DIFFERENCES BETWEEN *UNDERSTANDABLE STATISTICS* 9/E AND *UNDERSTANDING BASIC STATISTICS* 5/E vi

SPSS GUIDE

CHAPTER 1: GETTING STARTED

About the TI-83 Plus and TI-84 Plus Graphing Calculators 3

Using the TI-83 Plus and TI-84 Plus 3

Computations on the TI-83 Plus and TI-84 Plus.. 5

Entering Data.. 7

Comments on Entering and Correcting Data 9

Lab Activities to Get Started Using the TI-83 Plus and TI-84 Plus................................. 10

Random Samples 11

Lab Activities for Random Samples 13

CHAPTER 2: ORGANIZING DATA

Histograms 14

Lab Activities for Histograms 17

CHAPTER 3: AVERAGES AND VARIATION

One-Variable Statistics 19

Lab Activities for One-Variable Statistics... 22

Box-and-Whisker Plots.. 24

Lab Activities for Box-and-Whisker Plots 26

CHAPTER 4: CORRELATION AND REGRESSION

Linear Regression 28

Lab Activities for Linear Regression... 34

CHAPTER 5: ELEMENTARY PROBABILITY THEORY 36

CHAPTER 6: THE BINOMIAL PROBABILITY DISTRIBUTION AND RELATED TOPICS

Discrete Probability Distributions 37

Lab Activities for Discrete Probability Distributions 37

Binomial Probabilities 38

Lab Activities for Binomial Probabilities.. 40

CHAPTER 7: NORMAL DISTRIBUTIONS AND SAMPLING DISTRIBUTIONS

The Area Under Any Normal Curve 41

Lab Activities for the Area Under Any Normal Curve 43

Sampling Distributions and the Central Limit Theorem 44

CHAPTER 8: ESTIMATION

Confidence Intervals for a Population Mean 45

Lab Activities for Confidence Intervals for a Population Mean 47

Confidence Intervals for the Probability of Success p in a Binomial Distribution 47

Lab Activities for Confidence Intervals for the Probability of Success p in a Binomial Distribution 48

CHAPTER 9: HYPOTHESIS TESTING

Testing a Single Population Mean 49

Lab Activities for Testing a Single Population Mean 51

Testing Involving a Single Proportion 52

CHAPTER 10: INFERENCES ABOUT DIFFERENCES

Tests Involving Paired Differences (Dependent Samples) 53

Lab Activities Using Tests Involving Paired Differences (Dependent Samples) 55

Inferences About the Difference of Two Means $\mu_1 - \mu_2$ and the Difference of Two Proportions $p_1 - p_2$ 55

Lab Activities for Inferences About the Difference of Means (Independent Samples) and the Difference of Proportions 58

CHAPTER 11: ADDITIONAL TOPICS USING INFERENCE

Chi-Square Tests of Independence 60

Lab Activities for Chi-Square Test of Independence 63

APPENDIX

PREFACE A-3

SUGGESTIONS FOR USING THE DATA SETS A-4

DESCRIPTIONS OF DATA SETS A-5

Preface

The use of computing technology can greatly enhance a student's learning experience in statistics. *Understanding Basic Statistics* is accompanied by four Technology Guides, which provide basic instruction, examples, and lab activities for four different tools:

TI-83 Plus and TI-84 Plus

Microsoft Excel® 2007 with Analysis ToolPak for Windows®

MINITAB Version 15

SPSS Version 16

The TI-83 Plus and TI-84 Plus are versatile, widely available graphing calculators made by Texas Instruments. The calculator guide shows how to use their statistical functions, including plotting capabilities.

Excel is an all-purpose spreadsheet software package. The Excel guide shows how to use Excel's built-in statistical functions and how to produce some useful graphs. Excel is not designed to be a complete statistical software package. In many cases, macros can be created to produce special graphs, such as box-and-whisker plots. However, this guide only shows how to use the existing, built-in features. In most cases, the operations omitted from Excel are easily carried out on an ordinary calculator. The Analysis ToolPak is part of Excel and can be installed from the same source as the basic Excel program (normally, a CD-ROM) as an option on the installer program's list of Add-Ins. Details for getting started with the Analysis ToolPak are in Chapter 1 of this guide. No additional software is required to use the Excel functions described.

MINITAB is a statistics software package suitable for solving problems. It can be packaged with the text. Contact Cengage Learning for details regarding price and platform options.

SPSS is a powerful tool that can perform many statistical procedures. The SPSS guide shows how the manage data and perform various statistical procedures using this software.

The lab activities that follow accompany the text *Understanding Basic Statistics,* 5th edition by Brase and Brase. On the following page is a table to coordinate this guide with the parent text *Understandable Statistics,* 9th edition by Brase and Brase.

In addition, over one hundred data files from referenced sources are described in the Appendix. These data files are available via download at:

www.cengage.com/statistics/Brase/UBS5e

Understanding the Differences Between *Understandable Statistics* 9/e and *Understanding Basic Statistics* 5/e

Understandable Statistics is the full, two-semester introductory statistics textbook, which is now in its Ninth Edition.

Understanding Basic Statistics is the brief, one-semester version of the larger book. It is currently in its Fifth Edition.

Unlike other brief texts, *Understanding Basic Statistics* is not just the first six or seven chapters of the full text. Rather, topic coverage has been shortened in many cases and rearranged, so that the essential statistics concepts can be taught in one semester.

The major difference between the two tables of contents is that Regression and Correlation are covered much earlier in the brief textbook. In the full text, these topics are covered in Chapter 10. In the brief text, they are covered in Chapter 4.

Analysis of a Variance (ANOVA) is not covered in the brief text.

Understanding Statistics has 12 chapters and *Understanding Basic Statistics* has 11. The full text is a hardcover book, while the brief is softcover.

The same pedagogical elements are used throughout both texts.

The same supplements package is shared by both texts.

Following are the two Tables of Contents, side-by-side:

	Understandable Statistics (full)	*Understanding Basic Statistics* (brief)
Chapter 1	Getting Started	Getting Started
Chapter 2	Organizing Data	Organizing Data
Chapter 3	Averages and Variation	Averages and Variation
Chapter 4	Elementary Probability Theory	Correlation and Regression
Chapter 5	The Binomial Probability Distribution and Related Topics	Elementary Probability Theory
Chapter 6	Normal Distributions	The Binomial Probability Distribution and Related Topics
Chapter 7	Introduction to Sampling Distributions	Normal Curves and Sampling Distributions
Chapter 8	Estimation	Estimation
Chapter 9	Hypothesis Testing	Hypothesis Testing
Chapter 10	Correlation and Regression	Inferences About Differences
Chapter 11	Chi-Square and F Distributions	Additional Topics Using Inference
Chapter 12	Nonparametric Statistics	

SPSS v. 16 Guide

CHAPTER 1: GETTING STARTED

ABOUT THE TI-83 PLUS AND TI-84 PLUS GRAPHING CALCULATORS

Calculators with built-in statistical support provide tremendous aid in performing the calculations required in statistical analysis. The Texas Instruments TI-83 Plus and TI-84 Plus graphing calculators have many features that are particularly useful in an introductory statistics course. Among the features are:

(a) Data entry in a spreadsheet-like format

The TI-83 Plus and TI-84 Plus have six columns (called lists: L_1, L_2, L_3, L_4, L_5, and L_6) in which data can be entered. The data in a list can be edited, and new lists can be created using arithmetic that uses existing lists.

Sample Data Screen

L1	L2	L3	1
11.2	8.1	1	
8.7	6.2	2	
6.8	3.6	-1	
3.2	.9	2	
4.5	1	1	
------	------	1	
		5	

L1(1)=11.2

(b) Single-variable statistics: mean, standard deviation, median, maximum, minimum, quartiles 1 and 3, sums

(c) Graphs for single-variable statistics: histograms, box-and-whisker plots

(d) Estimation: confidence intervals using the normal distribution, or Student's t distribution, for a single mean and for a difference of means; confidence intervals for a single proportion and for the difference of two proportions

(e) Hypothesis testing: single mean (z or t); difference of means (z or t); proportions, difference of proportions, chi-square test of independence; linear regression

(h) Two-variable statistics: linear regression

(i) Graphs for two-variable statistics: scatter diagrams, graph of the least-squares line

In this *Guide* we will show how to use many features on the TI-83 Plus and TI-84 Plus graphing calculators to aid you as you study particular concepts in statistics. **Lab Activities** coordinated to the text *Understanding Basic Statistics, Fifth Edition,* are also included.

USING THE TI-83 PLUS AND TI-84 PLUS

The TI-83 Plus and TI-84 Plus have several functions associated with each key, and both calculators use the same keystroke sequences to perform those functions. With the TI-83 Plus, all the yellow items on the keypad are accessed by first pressing the yellow [2nd] key. On the TI-84 Plus these items are blue in color. For both calculators, the items in green are accessed by first pressing the green [ALPHA] key. The four arrow keys ▼▲▶◀ enable you to move through a menu screen or along a graph. On the following page is the image of a TI-84 Plus calculator. The keypad of the TI-83 Plus is exactly the same as that of the TI-84 Plus.

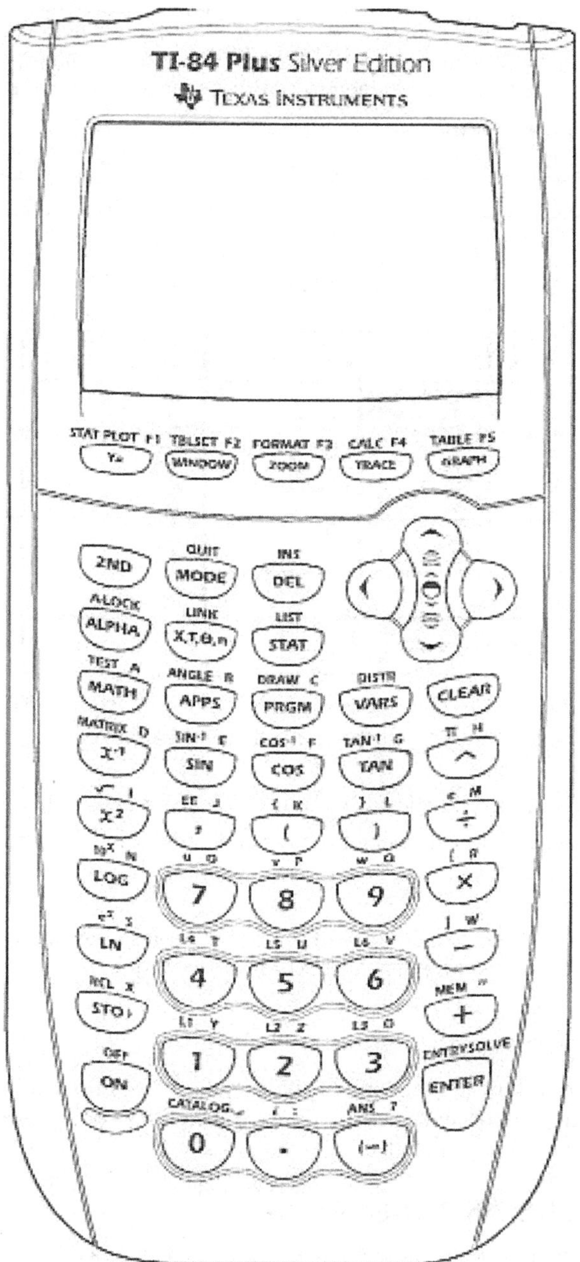

The calculator uses screen menus to access additional operations. For instance, to access the statistics menu, press [STAT]. Your screen should look like this:

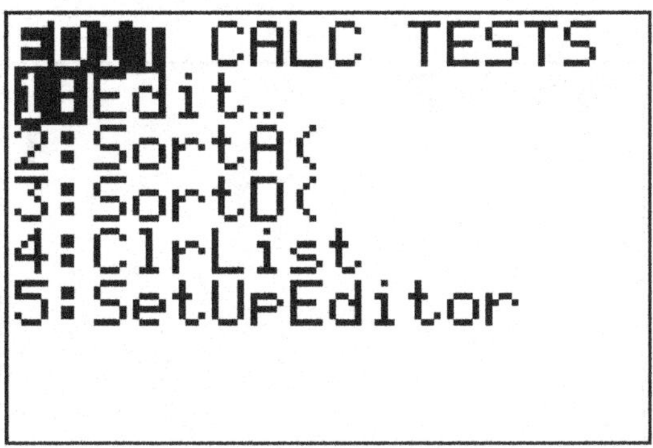

Use the arrow keys to highlight your selection. Pressing [ENTER] selects the highlighted item.

To leave a screen, press either [2nd] **[QUIT]** or [CLEAR], or select another menu from the keypad.

Now press [MODE]. When you use your calculator for statistics, the most convenient settings are as shown. (The TI-83 Plus does not have the **SET CLOCK** line; otherwise, it is identical to the TI-84 Plus.)

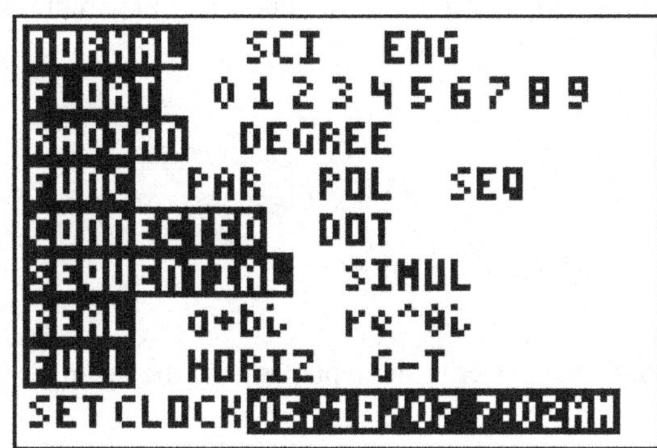

You can enter the settings by using the arrow keys to highlight selections and then pressing [ENTER].

COMPUTATIONS ON THE TI-83 PLUS AND TI-84 PLUS

In statistics, you will be evaluating a variety of expressions. The following examples demonstrate some basic keystroke patterns.

Example

Evaluate −2(3) + 7.

Use the following keystrokes: [(-)][2][×][3][+][7][ENTER] The result is 1.

To enter a negative number, be sure to use the key $\boxed{(-)}$ rather than the subtract key. Notice that the expression **-2*3 + 7** appears on the screen. When you press $\boxed{\text{ENTER}}$, the result **1** is shown on the far right side of the screen.

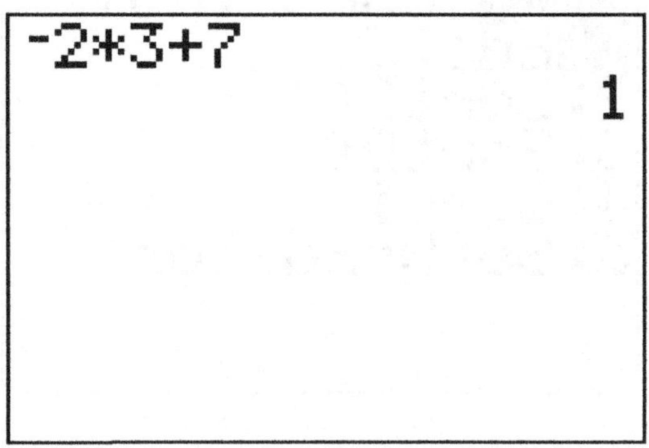

Example

(a) Evaluate $\dfrac{10-7}{3.1}$ and round the answer to three places after the decimal.

A reliable approach to evaluating fractions is to enclose the numerator in parentheses, and if the denominator contains more than a single number, enclose it in parentheses as well.

$$\frac{10-7}{3.1} = (10-7) \div 3.1$$

Use the keystrokes $\boxed{(}\boxed{1}\boxed{0}\boxed{-}\boxed{7}\boxed{)}\boxed{\div}\boxed{3}\boxed{.}\boxed{1}\boxed{\text{ENTER}}$.

The result is .9677419355, which rounds to .968.

(b) Evaluate $\dfrac{10-7}{\frac{3.1}{2}}$ and round the answer to three places after the decimal.

Place both numerator and denominator in parentheses: $(10-7) \div (3.1 \div 2)$

Use the keystrokes $\boxed{(}\boxed{1}\boxed{0}\boxed{-}\boxed{7}\boxed{)}\boxed{\div}\boxed{(}\boxed{3}\boxed{.}\boxed{1}\boxed{\div}\boxed{2}\boxed{)}\boxed{\text{ENTER}}$.

The result is 1.935483871, or 1.935 rounded to three places after the decimal.

Example

Several formulas in statistics require that we take the square root of a value. Note that a left parenthesis, (, is automatically placed next to the square root symbol when $\boxed{\text{2nd}}\ \boxed{\sqrt{}}$ is pressed. Be sure to close the parentheses after typing in the radicand.

(a) Evaluate $\sqrt{10}$ and round the result to three places after the decimal.

$\boxed{\text{2nd}}\boxed{\sqrt{}}\boxed{1}\boxed{0}\boxed{)}\boxed{\text{ENTER}}$

The result is 3.16227766, which rounds to 3.162.

(b) Evaluate $\frac{\sqrt{10}}{3}$ and round the result to three places after the decimal.

[2nd][√][1][0][)][÷][3][ENTER]

The result rounds to 1.054.

Be careful to close the parentheses. If you do not close the parentheses, you will get the result of $\sqrt{\frac{10}{3}} \approx 1.826$.

Example

Some expressions require us to use powers.

(a) Evaluate 3.2^2.

In this case, we can use the [x^2] key.

[3][.][2][x^2][ENTER]

The result is 10.24.

(b) Evaluate 0.4^3.

In this case, we use the [^] key.

[.][4][^][3][ENTER]

The result is .064.

ENTERING DATA

To use the statistical processes built into the TI-83 Plus and TI-84 Plus, we first enter data into lists.

Press the [STAT] key. Next we will clear any existing data lists. With **EDIT** highlighted, select **4:ClrList** using the arrow keys and press [ENTER].

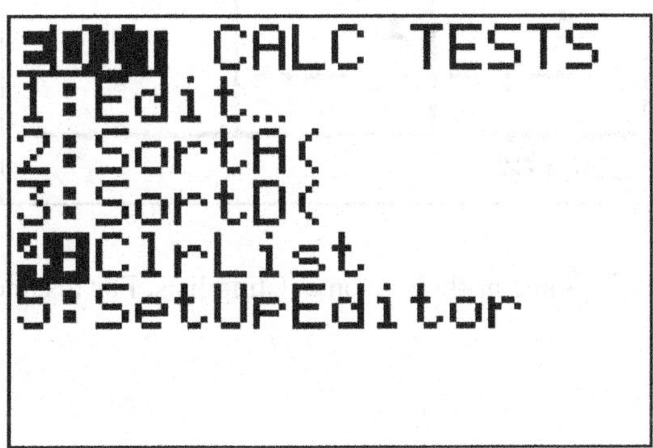

Then type in the six lists separated by commas as shown. (The lists are found using the [2nd] key, followed by L_1, L_2, L_3, L_4, L_5, and L_6 .) Press [ENTER].

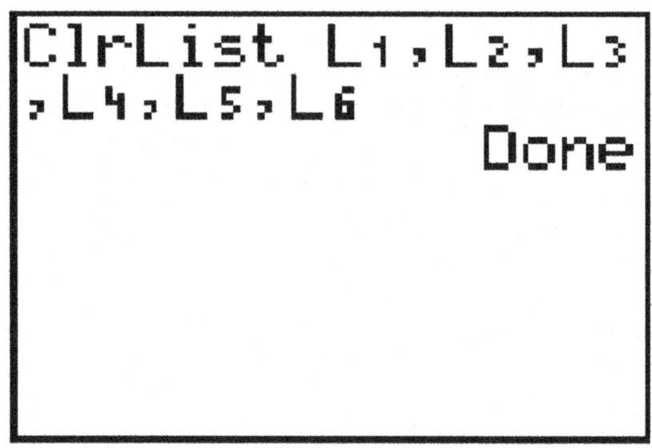

Next press [STAT] again and select item **1:Edit...**. You will see the data screen and are now ready to enter data.

Let's enter the numbers.

    2   5   7   9   11   in list L_1

    3   6   9 11   1   in list L_2

Press [ENTER] after each number, and use the arrow keys to move between L_1 and L_2.

To correct a data entry error, highlight the entry that is wrong and enter the correct data value.

We can also create new lists by doing arithmetic on existing lists. For instance, let's create L_3 by using the formula $L_3 = 2L_1 + L_2$.

Highlight L_3. Then type in $\mathbf{2L_1 + L_2}$ and press $\boxed{\text{ENTER}}$.

The final result is shown below.

To leave the data screen, press $\boxed{\text{2nd}}$ **[QUIT]** or press another menu key, such as $\boxed{\text{STAT}}$.

COMMENTS ON ENTERING AND CORRECTING DATA

(a)   To create a new list from old lists, the lists must all have the same number of entries.

(b)   To delete a data entry, highlight the data value you wish to correct and press the $\boxed{\text{DEL}}$ key.

(c)   To insert a new entry into a list, highlight the position directly below the place you wish to insert the new value. Press $\boxed{\text{2nd}}$ **[INS]** and then enter the data value.

LAB ACTIVITIES TO GET STARTED USING THE TI-83 PLUS AND TI-84 PLUS

1. Practice doing some calculations using the TI-83 Plus and TI-84 Plus. Be sure to use parentheses around a numerator or denominator that contains more than a single number. Round all answers to three places after the decimal.

   (a) $\dfrac{5-2.3}{1.3}$ Ans. 2.077   **(b)** $\dfrac{8-3.3}{\frac{3}{2}}$ Ans. 3.133

   (c) $-2(3.4)+5.8$ Ans. -1 **(d)** $-4(-1.7)-2.1$ Ans. 4.7

   (e) $\sqrt{5.3}$ Ans. 2.302   **(f)** $\sqrt{6+3(2)}$ Ans. 3.464

   (g) $\sqrt{\dfrac{8-2.7}{5-1}}$ Ans. 1.151   **(h)** 1.5^2 Ans. 2.25

   (i) $(5-7.2)^2$ Ans. 4.84   **(j)** $(0.7)^3(0.3)^2$ Ans. 0.031

2. Enter the following data into the designated list. Be sure to clear all lists first.

   L_1:   3   7   9.2   12 -4

   L_2: -2 9   4.3   16   10

   L_3: $L_1 - 2$

   L_4: $-2L_2 - L_1$

   L_1: Change the data value in the second position to 6.

RANDOM SAMPLES (SECTION 1.2 OF *UNDERSTANDING BASIC STATISTICS*)

The TI-83 Plus and TI-84 Plus graphics calculators have a random number generator, which can be used in place of a random number table. Press MATH. Use the arrow keys to highlight **PRB**. Notice that **1:rand** is selected.

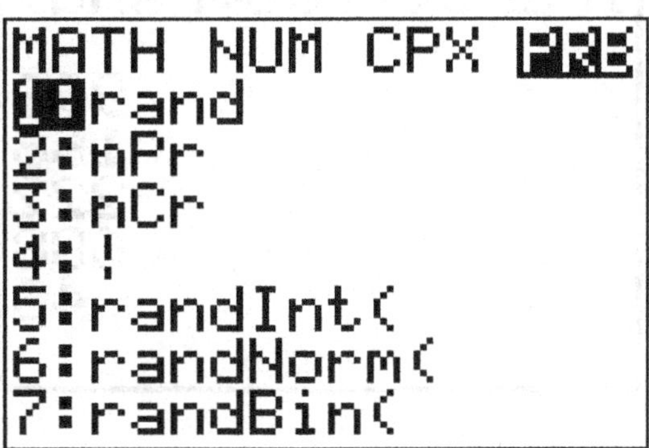

When you press ENTER, rand appears on the screen. Press ENTER again and a random number between 0 and 1 appears.

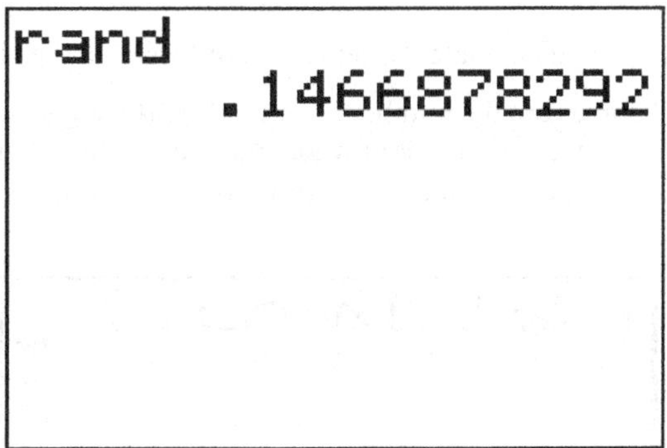

To generate a random whole number with up to 3 digits, multiply the output of **rand** by 1000 and take the integer part. The integer-part command is found in the MATH menu under **NUM,** selection **3:iPart (**. Press ENTER to display the command on the main screen.

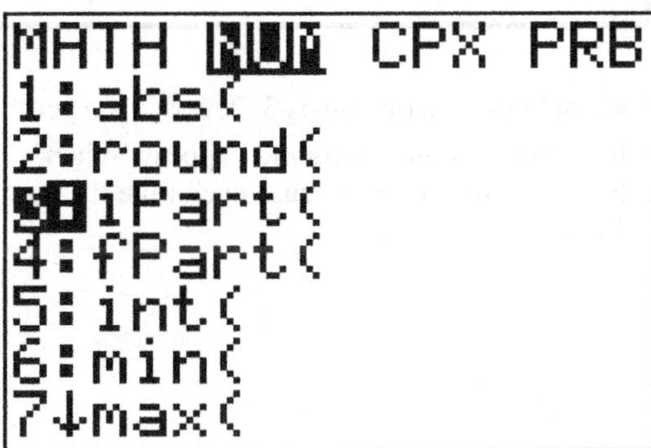

Now let's put all commands together to generate random whole numbers with up to 3 digits. First display **iPart(**. Then enter **1000*rand** and close the parentheses. Now, each time you press ENTER, a new random number will appear. Notice that the numbers might be repeated.

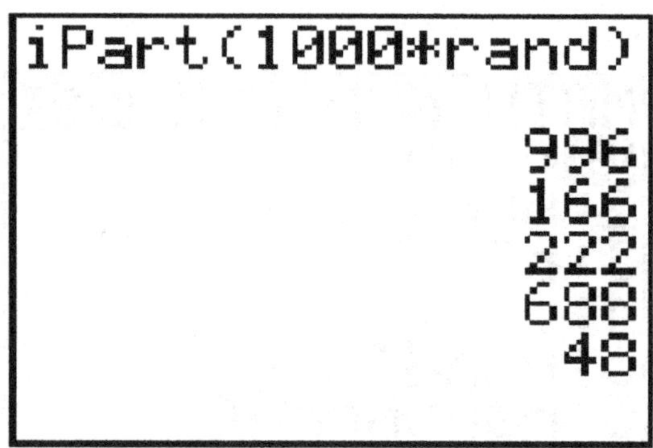

To generate random whole numbers with up to 2 digits, multiply **rand** by 100 instead of 1000. For random numbers with 1 digit, multiply **rand** by 10.

Simulating experiments in which outcomes are equally likely is an important use of random numbers.

Example

Use the TI-83 Plus or TI-84 Plus to simulate the outcomes of tossing a die six times. Record the results.

In this case, the possible outcomes are the numbers 1 through 6. If we generate a random number outside of this range we will ignore it. Because we want random whole numbers with 1 digit, we will follow the method prescribed earlier and multiply each random number by 10. Press ENTER 6 times.

Of the first six values listed, we only use the outcomes 3, 2, and 2. To get the remaining outcomes, keep pressing ENTER until three more digits in the range 1 through 6 appear. When we did this, the next three such digits were 1, 4, and 3. Your results will be different, because each time the random number generator is used, it gives a different sequence of numbers.

LAB ACTIVITIES FOR RANDOM SAMPLES

1. Out of a population of 800 eligible county residents, select a random sample of 20 people for prospective jury duty. By what value should you multiply **rand** to generate random whole numbers with 3 digits? List the first 20 numbers corresponding to people for prospective jury duty.

2. We can simulate dealing bridge hands by numbering the cards in a bridge deck from 1 to 52. Then we draw a random sample of 13 numbers without replacement from the population of 52 numbers. A bridge deck has 4 suits: hearts, diamonds, clubs, and spades. Each suit contains 13 cards: those numbered 2 through 10, a jack, a queen, a king, and an ace. Decide how to assign the numbers 1 through 52 to the cards in the deck. Use the random number generator on the TI-83 Plus or TI-84 Plus to get the numbers of the 13 cards in one hand. Translate the numbers to specific cards and tell what cards are in the hand. For a second game, the cards would be collected and reshuffled. Using random numbers from the TI-83 Plus or TI-84 Plus, determine the hand you might get in a second game.

CHAPTER 2: ORGANIZING DATA

HISTOGRAMS (SECTION 2.1 OF *UNDERSTANDING BASIC STATISTICS*)

The TI-83 Plus and TI-84 Plus graphing calculators draw histograms for data entered in the data lists. The calculators follow the convention that if a data value falls on a class limit, then it is tallied in the frequency of the bar to the right. However, if you specify the lower class boundary of the first class and the class width (see *Understanding Basic Statistics* for procedures to find these values), then no data will fall on a class boundary. The following example shows you how to draw histograms.

Example

Throughout the day from 8 A.M. to 11 P.M., Tiffany counted the number of ads occurring every hour on one commercial TV station. The 15 data values are as follows:

$$10 \quad 12 \quad 8 \quad 7 \quad 15 \quad 6 \quad 5 \quad 8$$

$$8 \quad 10 \quad 11 \quad 13 \quad 15 \quad 8 \quad 9$$

To make a histogram with 4 classes, first enter the data in L_1.

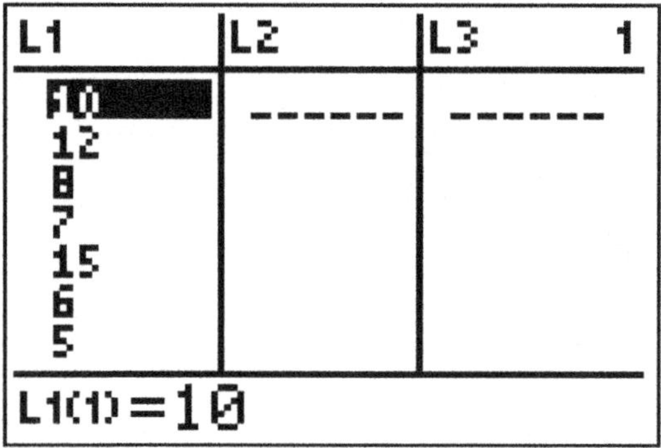

Next, we will set the graphing window. However, we need to know both the lower class boundary for the first class and the class width. Use the techniques in Section 2.1 of *Understanding Basic Statistics* to find these values.

The smallest data value is 5, so the lower class boundary of the first class is 4.5. The class width is

$$\frac{\text{largest data value} - \text{smallest data value}}{\text{Number of classes}}, \textit{ increased to the next integer}$$

$$\frac{15-5}{4} = 2.5, \textit{ increased to } 3$$

Now we set the graphing window. Press the WINDOW key.

Select **Xmin** and enter the *lower class boundary* of the first class. For this example, **Xmin = 4.5**.

Select **Xscl** and enter the *class width*. For this example, **Xscl = 3**.

Setting **Xmax = 17** ensures that the histogram fits on the screen, as our highest data value is 15.

Setting **Ymin = -5** and **Ymax = 15** makes sure the histogram fits comfortably on the screen.

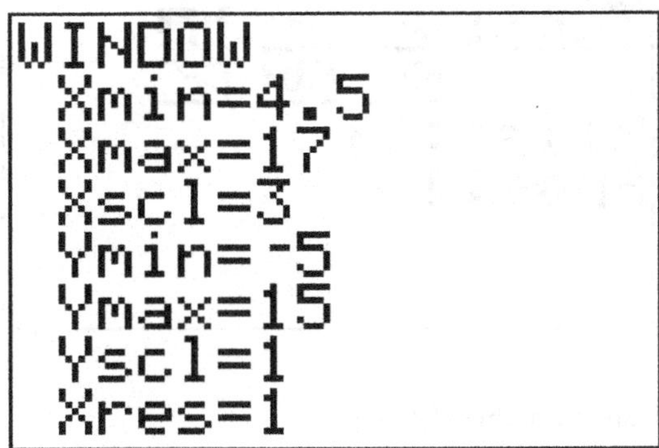

The next step is to select histogram style for the plot. Press 2nd [STAT PLOT].

Highlight **1:Plot1...** and press ENTER.

Then on the next screen select **On** and the histogram shape.

In this example we stored the data in L_1. Set **Xlist** to L_1 by pressing [2nd] [L_1]. Each data value is listed once for each time it occurs, so **1** is entered for **Freq**.

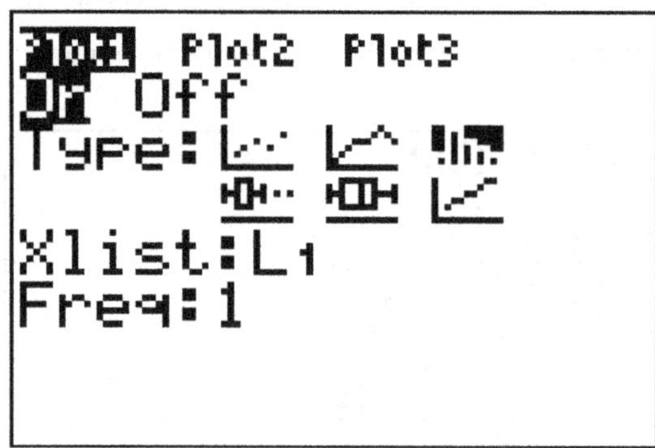

We are ready to graph the histogram. Press [GRAPH].

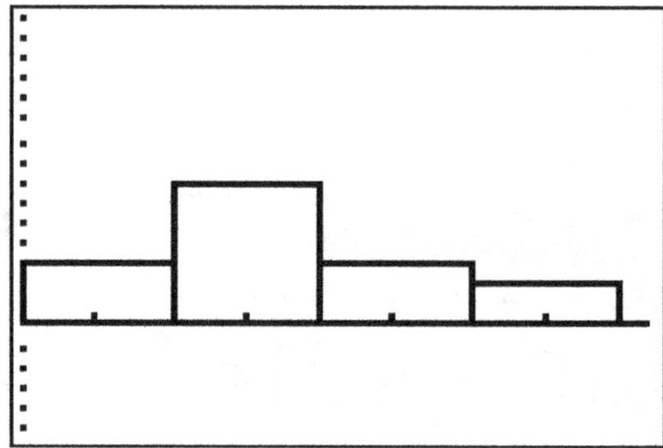

Press [TRACE] and notice that a blinking cursor appears over the first bar. The class boundaries of the bar are given as **min** and **max.** The frequency is given by **n.** Use the right and left arrow keys to move from bar to bar.

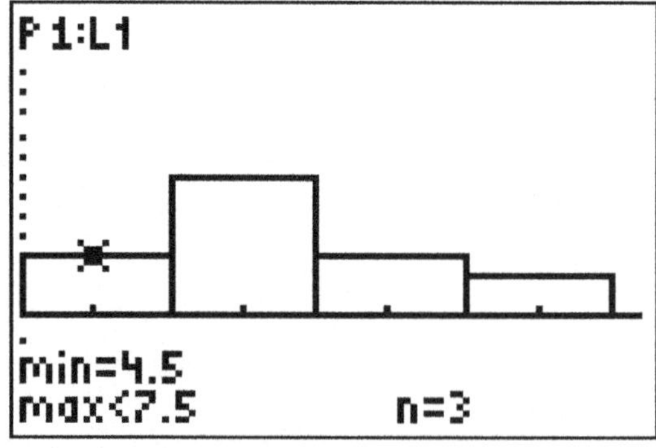

After you have graphed the histogram, you can use [TRACE] and the arrow keys to find the class boundaries and frequency of each class. The calculator has done the work of sorting and tallying the data for each class.

LAB ACTIVITIES FOR HISTOGRAMS

1. A random sample of 50 professional football players produced the following data. (See **Weights of Pro Football Players** (data file Sv02.txt).)

   The following data represent weights in pounds of 50 randomly selected pro football linebackers.

   Source: The Sports Encyclopedia Pro Football, 1960–1992

225	230	235	238	232	227	244	222
250	226	242	253	251	225	229	247
239	223	233	222	243	237	230	240
255	230	245	240	235	252	245	231
235	234	248	242	238	240	240	240
235	244	247	250	236	246	243	255
241	245						

   Enter these data in L_1. Scan the data for the low and high values. Compute the class width for 5 classes and find the lower class boundary for the first class. Use the TI-83 Plus or TI-84 Plus to make a histogram with 5 classes. Repeat the process for 9 classes. Is the data distribution skewed or symmetric? Is the distribution shape more pronounced with 5 classes or with 9 classes?

2. Explore the other data files found in the Appendix of this *Guide*, such as

   Disney Stock Volume (data file Sv01.txt)

   Heights of Pro Basketball Players (data file Sv03.txt)

   Miles per Gallon Gasoline Consumption (data file Sv04.txt)

   Fasting Glucose Blood Tests (data file Sv05.txt)

   Number of Children in Rural Canadian Families (data file Sv06.txt)

   Select one of these files and make a histogram with 7 classes. Comment on the histogram shape.

3. **(a)** Consider the following data:

1	3	7	8	10
6	5	4	2	1
9	3	4	5	2

   Place the data in L_1. Use the TI-83 Plus or TI-84 Plus to make a histogram with $n = 3$ classes. Jot down the class boundaries and frequencies so that you can compare them to part (b).

   (b) Now add 20 to each data value of part (a). The results are as follows:

21	23	27	28	30
26	25	24	22	21
29	23	24	25	22

   By using arithmetic in the data list screen, you can create L_2 and let $L_2 = L_1 + 20$.

Make a histogram with 3 classes. Compare the class boundaries and frequencies with those obtained in part (a). Are each of the boundary values 20 more than those of part (a)? How do the frequencies compare?

(c) Use your discoveries from part (b) to predict the class boundaries and class frequency with 3 classes for the data values below. What do you suppose the histogram will look like?

1001	1003	1007	1008	1010
1006	1005	1004	1002	1001
1009	1003	1004	1005	1002

Would it be safe to say that we simply shift the histogram of part (a) 1000 units to the right?

(d) What if we multiply each of the values of part (a) by 10? Will we effectively multiply the entries for using class boundaries by 10? To explore this question create a new list L_3 using the formula $L_3 = 10L_1$. The entries will be as follows:

10	30	70	80	100
60	50	40	20	10
90	30	40	50	20

Compare the histogram with three classes to the one from part (a). You will see that there does not seem to be an exact correspondence. To see why, look at the class width and compare it to the class width of part (a). The class width is always increased to the next integer value no matter how large the integer data values are. Consequently, the class width for the data in part (d) was increased to 31 rather than 40.

4.   Histograms are not effective displays for some data. Consider the data

1	2	3	6	5	7
9	8	4	12	11	15
14	12	6	2	1	206

Use the TI-83 Plus or TI-84 Plus to make a histogram with 2 classes. Then change to 3 classes, on up to 10 classes. Notice that all the histograms lump the first 17 data values into the first class, and the one data value, 206, in the last class. What feature of the data causes this phenomenon? Recall that

$$\text{Class width} = \frac{\text{largest value} - \text{smallest value}}{\text{number of class}}, \text{ increased to the next integer.}$$

How many classes would you need before you began to see the first 17 data values distributed among several classes? What would happen if you simply did not include the extreme value 206 in your histogram?

CHAPTER 3: AVERAGES AND VARIATION

ONE-VARIABLE STATISTICS (SECTIONS 3.1 AND 3.2 OF *UNDERSTANDING BASIC STATISTICS*)

The TI-83 Plus and TI-84 Plus graphing calculators support many of the common descriptive measures for a data set. The measured supported are

Mean $\bar{x}$

Sample standard deviations S_x

Population standard deviation σ_x

Number of data values n

Median

Minimum data value

Maximum data value

Quartile 1

Quartile 3

$\sum x$

$\sum x^2$

Although the mode is not provided directly, the menu choice **SortA** sorts the data in ascending order. By scanning the sorted list, you can find the mode fairly quickly.

Example

At Lazy River College, 15 students were selected at random from a group registering on the last day of registration. The times (in hours) necessary for these students to complete registration follow:

1.7	2.1	0.8	3.5	1.5	2.6	2.1	2.08
3.1	2.1	1.3	0.5	2.1	1.5	1.9	

Use the TI-83 Plus or TI-84 Plus to find the mean, sample standard deviation, and median. Use the **SortA** command to sort the data and scan for the mode, if it exists.

First enter the data into list L_1.

Press $\boxed{\text{STAT}}$ again, and highlight **CALC** with item **1:1-Var Stats**.

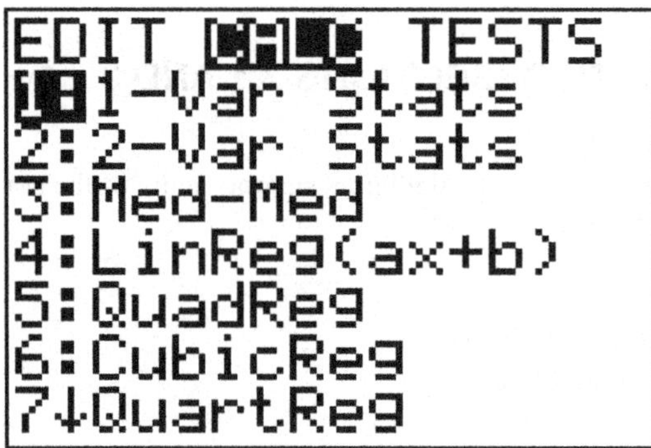

Press $\boxed{\text{ENTER}}$. The command **1-Var Stats** will appear on the screen. Then tell the calculator to use the data in list L_1 by pressing $\boxed{\text{2nd}}$ $[\mathbf{L_1}]$.

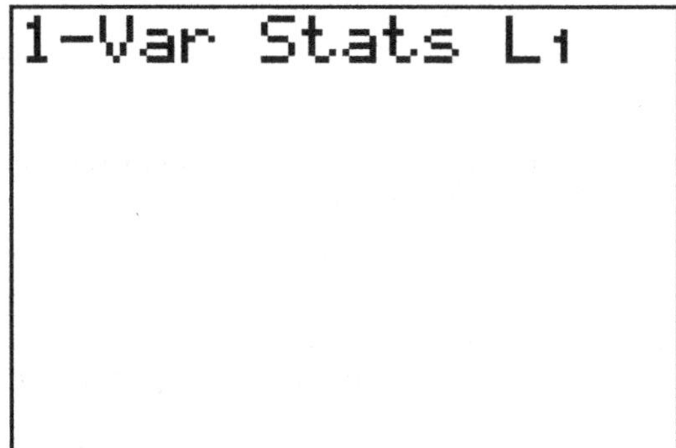

Press $\boxed{\text{ENTER}}$. The next two screens contain the statistics for the data in List L_1 .

Note the arrow $\downarrow$ on the last line. This is an indication that you may use the down-arrow key to display more results.

Scroll down. The second full screen shows the rest of the statistics for the data in list L_1.

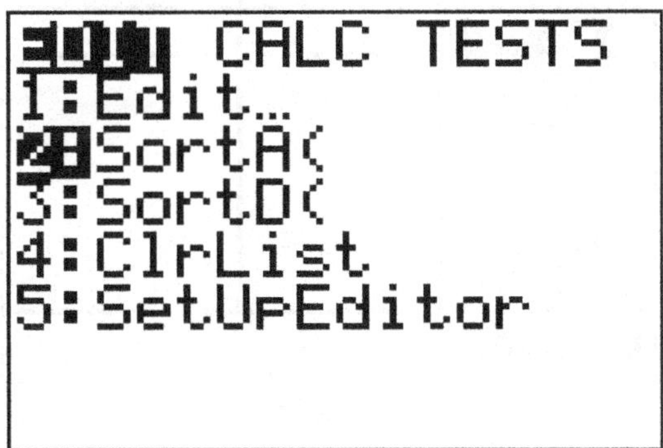

To sort the data, first press [STAT] to return to the edit menu. Then highlight **EDIT** and item **2:SortA(**.

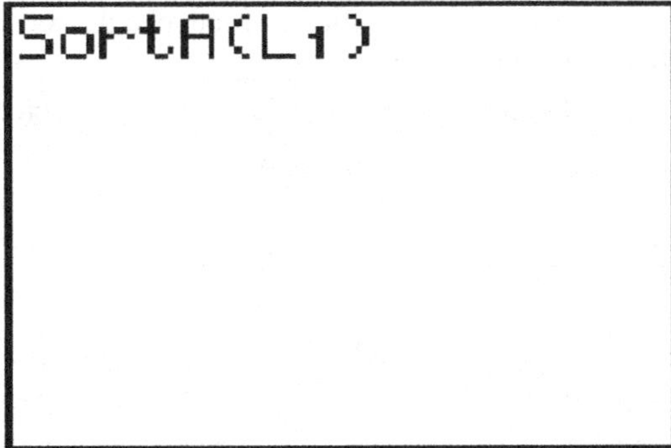

Press [ENTER]. The command **SortA** appears on the screen. Type in [2nd] **[L₁])**.

```
SortA(L₁)
```

When you press [ENTER], the comment **Done** appears on the screen.

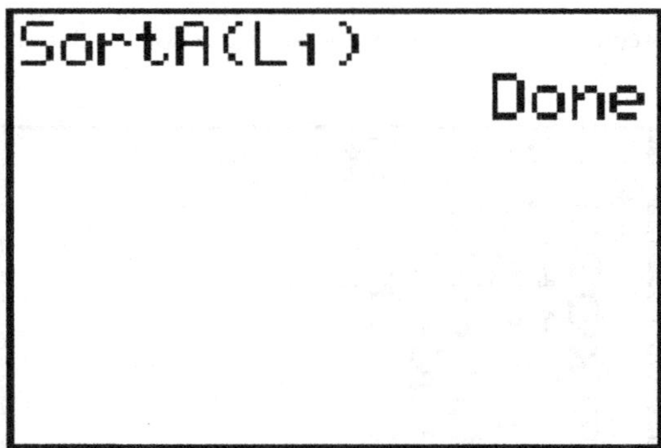

Press STAT and return to the data entry screen. Notice that the data in list L_1 are now sorted in ascending order.

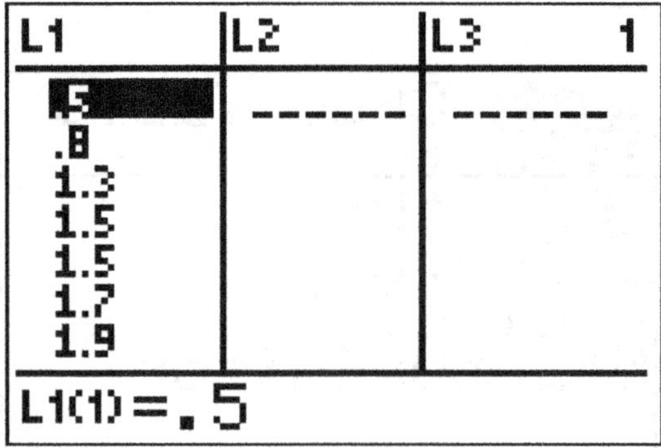

A scan of the data shows that the mode is 2.1, as this data value occurs more than any other.

LAB ACTIVITIES FOR ONE-VARIABLE STATISTICS

1. A random sample of 50 professional football players produced the following data. (See **Weights of Pro Football Players** (data file Sv02.txt).)

    The following data represent weights in pounds of 50 randomly selected pro football linebackers.

    Source: The Sports Encyclopedia Pro Football, 1960–1992

225	230	235	238	232	227	244	222
250	226	242	253	251	225	229	247
239	223	233	222	243	237	230	240
255	230	245	240	235	252	245	231
235	234	248	242	238	240	240	240
235	244	247	250	236	246	243	255
241	245						

    Enter these data in L_1. Use **1-Var Stats** to find the mean, median, and standard deviation for the weights. Use **SortA** to sort the data and scan for a mode if one exists.

2. Explore some of the other data sets found in the Appendix of this *Guide*, such as

  Disney Stock Volume (data file Sv01.txt)

  Heights of Pro Basketball Players (data file Sv03.txt)

  Miles per Gallon Gasoline Consumption (data file Sv04.txt)

  Fasting Glucose Blood Tests (data file Sv05.txt)

  Number of Children in Rural Canadian Families (data file Sv06.txt)

 Use the TI-83 Plus or TI-84 Plus to find the mean, median, standard deviation, and mode of the data.

3. In this problem we will explore the effects of changing data values by multiplying each data value by a constant, or by adding the same constant to each data value.

 (a) Consider the data

1	8	3	5	7
2	10	9	4	6
3	5	2	9	1

 Enter the data into list L_1 and use the TI-83 Plus or TI-84 Plus to find the mode (if it exists), mean, sample standard deviation, range, and median. Make a note of these values, as you will compare them to those obtained in parts (b) and (c).

 (b) Now multiply each data value of part (a) by 10 to obtain the data

10	80	30	50	70
20	100	90	40	60
30	50	20	90	10

 Remember, you can enter these data in L_2 by using the command $L_2 = 10L_1$.

 Again, use the TI-83 Plus or TI-84 Plus to find the mode (if it exists), mean, sample standard deviation, range, and median. Compare these results to the corresponding ones from part (a). Which values changed? Did those that changed change by a factor of 10? Did the range or standard deviation change? Referring to the formulas for these measures (see Section 3.2 of *Understanding Basic Statistics*), can you explain why the values behaved the way they did? Will these results generalize to the situation of multiplying each data entry by 12 instead of by 10? What about multiplying each by 0.5? Predict the corresponding values that would occur if we multiplied the data set of part (a) by 1000.

 (c) Now suppose we add 30 to each data value of part (a):

31	38	33	35	37
32	40	39	34	36
33	35	32	39	31

 To enter these data, create a new list L_3 by using the command $L_3 = L_1 + 30$.

 Again, use the TI-83 Plus or TI-84 Plus to find the mode (if it exists), mean, sample standard deviation, range, and median. Compare these results with the corresponding ones of part (a). Which values changed? Of those that are different, did each change by being 30 more than the corresponding value of part (a)? Again, look at the formulas for range and standard deviation. Can you predict the observed behavior from the formulas? Can you generalize these results? What if we added 50 to each data value of part (a)? Predict the values for the mode, mean, sample standard deviation, range, and median.

BOX-AND-WHISKER PLOTS (SECTION 3.3 OF *UNDERSTANDING BASIC STATISTICS*)

The box-and-whisker plot is based on the five-number summary values found under **1-Var Stats**:

Lowest value

Quartile 1, or Q_1

Median

Quartile 3, or Q_3

Highest value

Example

Let's make a box-and-whisker plot using the data about the time it takes to register for Lazy River College students who waited till the last day to register. The times (in hours) are

1.7	2.1	0.8	3.5	1.5	2.6	2.1	2.8
3.1	2.1	1.3	0.5	2.1	1.5	1.9	

In a previous example we found the **1-Var Stats** for these data. First enter the data into list L_1.

Press $\boxed{\text{WINDOW}}$ to set the graphing window. Set **Xmin** to 0.5, as that is the smallest value. Set **Xmax** to 3.5, as that is the largest value. For both **Xscl** and **Yscl** enter 1. Use **Ymin = -5** and **Ymax = 5** to position the axis in the window.

Press [2nd] [**STAT PLOT**] and highlight **1:Plot...**. Press [ENTER].

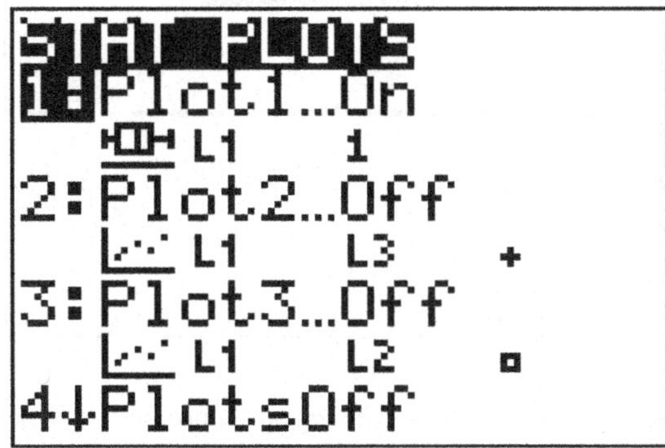

Highlight **On,** and select the box plot. The **Xlist** should be L_1 and **Freq** should be 1.

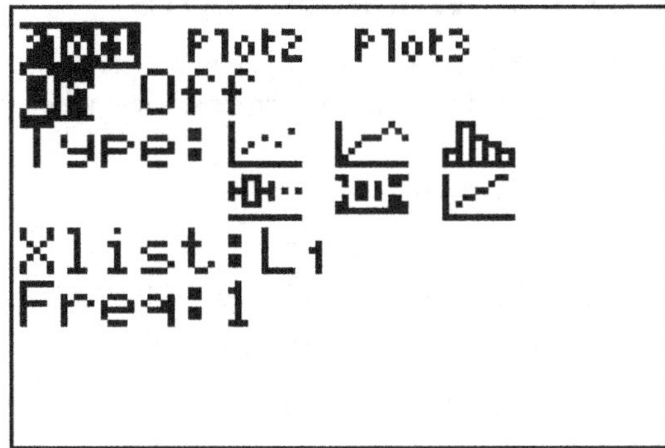

Press [GRAPH].

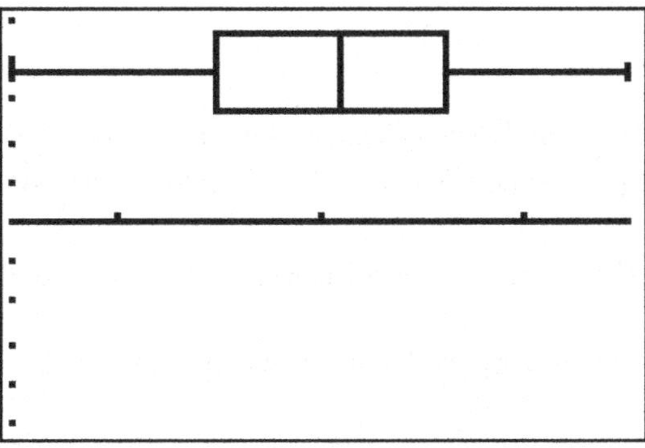

Press [TRACE] and use the arrow keys to display the five-number summary.

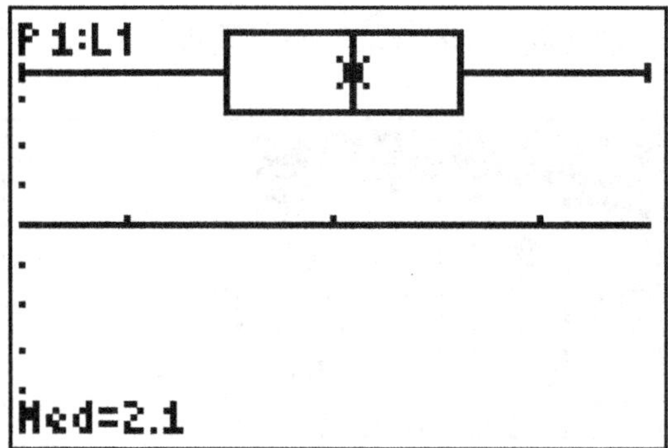

You now have many descriptive tools available: histograms, box-and-whisker plots, averages, and measures of variation such as the standard deviation. When you use all of these tools, you get a lot of information about the data.

LAB ACTIVITIES FOR BOX-AND-WHISKER PLOTS

1. One of the data sets included in the Appendix gives the miles per gallon gasoline consumption for a random sample of 55 makes and models of passenger cars. (See **Miles per Gallon Gasoline Consumption** (data file Sv04.txt)).

    Source: Environmental Protection Agency

30	27	22	25	24	25	24	15
35	35	33	52	49	10	27	18
20	23	24	25	30	24	24	24
18	20	25	27	24	32	29	27
24	27	26	25	24	28	33	30
13	13	21	28	37	35	32	33
29	31	28	28	25	29	31	

    Enter these data into list L_1, and then make a histogram and a box-and-whisker plot, and compute the mean, median, and sample standard deviation. Based on the information you obtain, respond to the following questions:

    (a) Is the distribution skewed or symmetric? How is this shown in both the histogram and the box-and-whisker plot?

    (b) Look at the box-and-whisker plot. Are the data more spread out above the median or below the median?

    (c) Look at the histogram and estimate the location of the mean on the horizontal axis. Are the data more spread out above the mean or below the mean?

    (d) Do there seem to be any data values that are unusually high or unusually low? If so, how do these show up on a histogram or on a box-and-whisker plot?

    (e) Pretend that you are writing a brief article for a newspaper. Describe the information about the data in non-technical terms. Be sure to make some comments about the "average" of the data values and some comments about the spread of the data.

2. Consider the test scores of 30 students in a political science class.

85	73	43	86	73	59	73	84	62	100
75	87	70	84	97	62	76	89	90	83
70	65	77	90	84	80	68	91	67	79

(a) For this population of test scores, find the mode, median, mean, range, variance, standard deviation, coefficient of variation, and the five-number summary. Then make a box-and-whisker plot. Be sure to record all of these values so you can compare them to the results of part (b).

(b) Greg was a student in the political science class of part (a). Suppose he missed a number of classes because of illness, but took the exam anyway and made a score of 30 instead of 85 as listed as the first entry of the data in part (a). Again, use the calculator to find the mode, median, mean, range, variance, standard deviation, coefficient of variation, and the five-number summary of the revised data. Make a box-and-whisker plot using the new data set. Compare these results to the corresponding results of part (a). Which average was most affected: mode, median, or mean? What about the range, standard deviation, and coefficient of variation? How do the box-and-whisker plots compare?

(c) Write a brief essay in which you use the results of parts (a) and (b) to predict how an extreme data value affects a data distribution. What do you predict for the results if Greg's test score had been 80 instead of 30 or 85?

CHAPTER 4: CORRELATION AND REGRESSION

LINEAR REGRESSION (SECTIONS 4.1 AND 4.2 OF *UNDERSTANDING BASIC STATISTICS*)

Important Note: Before beginning this chapter, press 2nd **[CATALOG] (above** 0 **) and scroll down to the entry DiagnosticOn. Press** ENTER **twice.** After doing this, the regression correlation coefficient r will appear as output with the linear regression line.

The TI-83 Plus and TI-84 Plus graphing calculators have automatic regression functions built-in as well as summary statistics for two variables. To locate the regression menu, press STAT, and then select **CALC**. The choice **8:LinReg(a+bx)** performs linear regression on the variables in L_1 and L_2. If the data are in other lists, then specify those lists after the **LinReg(a+bx)** command.

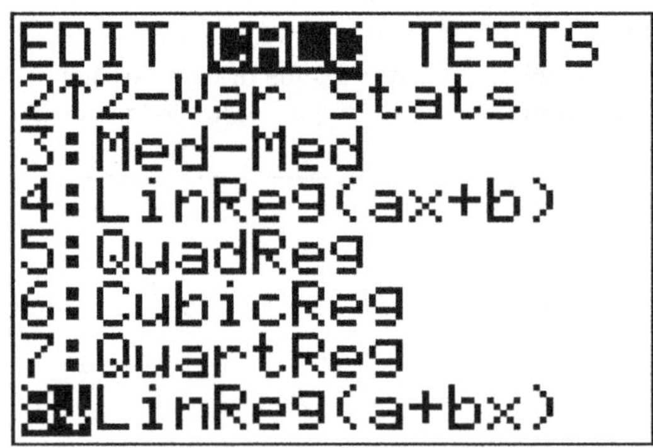

Let's look at a specific example to see how to do linear regression on the TI-83 Plus or TI-84 Plus.

Example

Merchandise loss due to shoplifting, damage, and other causes is called shrinkage. Shrinkage is a major concern to retailers. The managers of H. R. Merchandise believe there is a relationship between shrinkage and number of clerks on duty. To explore this relationship, a random sample of 7 weeks was selected. During each week the staffing level of sales clerks was kept constant and the dollar value of the shrinkage was recorded.

Number of Sales Clerks:	12	11	15	9	13	8
Shrinkage:	15	20	9	25	12	31

(a) Find the equation of the least-squares line, with number of sales clerks as the explanatory variable. First put the numbers of sales clerks in L_1 and the corresponding amount of shrinkage in L_2.

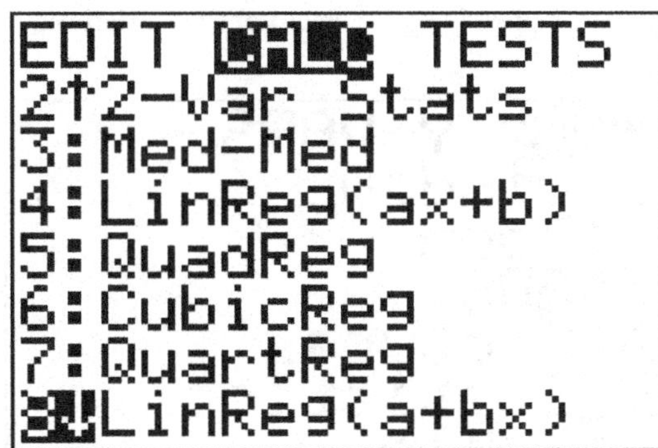

Use the **CALC** menu and select **8:LinReg(a+bx)**.

Designate the lists containing the data. Then press ENTER.

```
LinReg(a+bx) L₁,
L₂
```

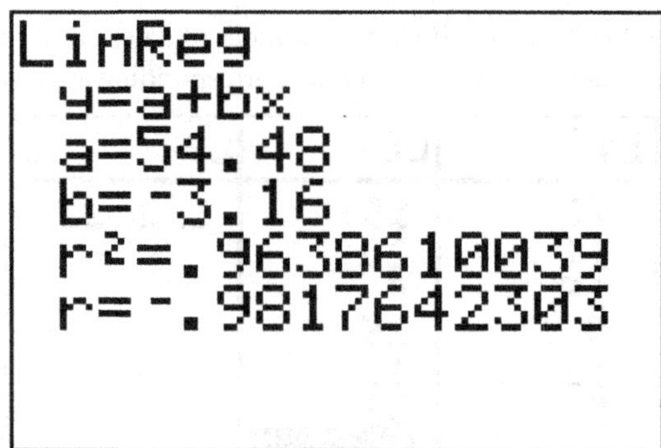

Note that the linear regression equation is $y = 54.48x - 3.16x$. We also have the value of the Pearson correlation coefficient, r and the coefficient of determination, r^2.

(b) Make a scatter diagram and show the least-squares line on the plot.

To graph the least squares line, press Y=. Clear all existing equations from the display. To enter the least-squares equation directly, press VARS. Select item **5:Statistics...**.

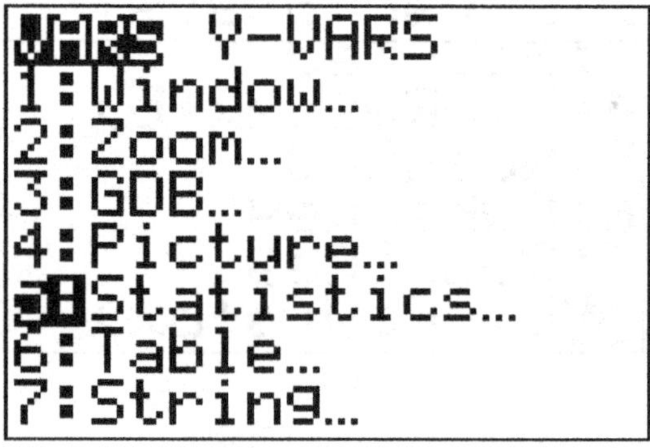

Press ENTER and then select **EQ**, and then **1:RegEQ**.

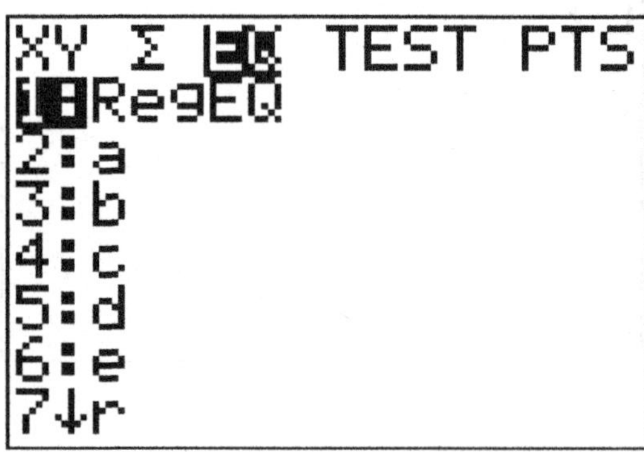

Press [ENTER]. You will see the least-squares regression equation appear automatically in the equations list.

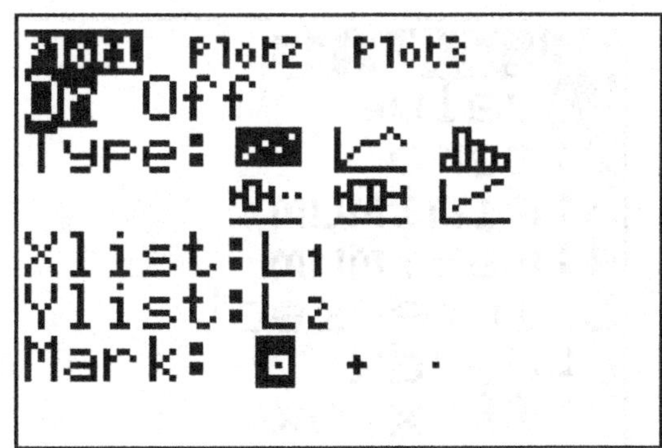

We want to graph the least-squares line with the scatter diagram. Press [2nd] [STAT PLOT]. Select **1:Plot1...**. Then select **On**, highlight the scatter diagram picture, set **Xlist** to L_1, and set **Ylist** to L_2.

The last step before we graph is to set the graphing window. Press [WINDOW]. Set **Xmin** to the lowest x value , and set **Xmax** to the highest x value. For **Xscl** choose 1. Use **Ymin = -5** and **Ymax = 30**, and set **Yscl=5**.

Now press GRAPH.

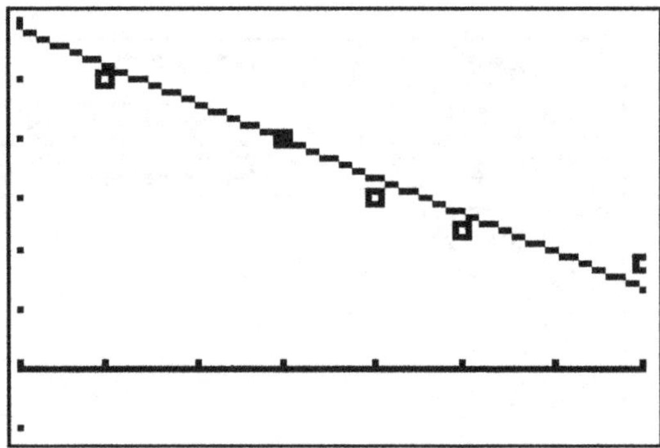

(c) Predict the shrinkage when 10 clerks are on duty.

Press 2nd **[CALC]**. This is the Calculate function for graphs and is different from the CALC you highlight from the STAT menu. Highlight **1:value**.

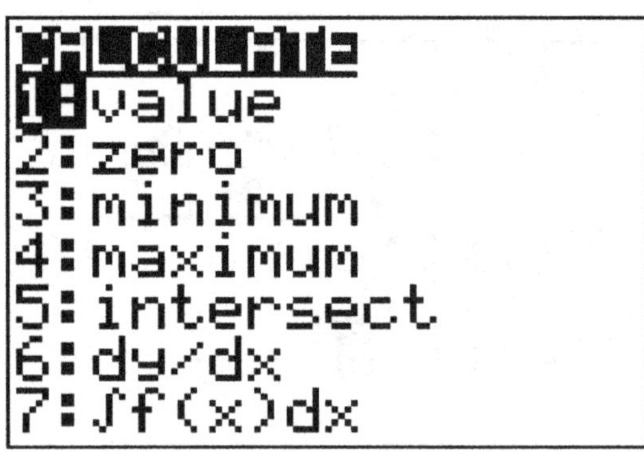

Press ENTER. The graph will appear. Enter 10 next to **X=**.

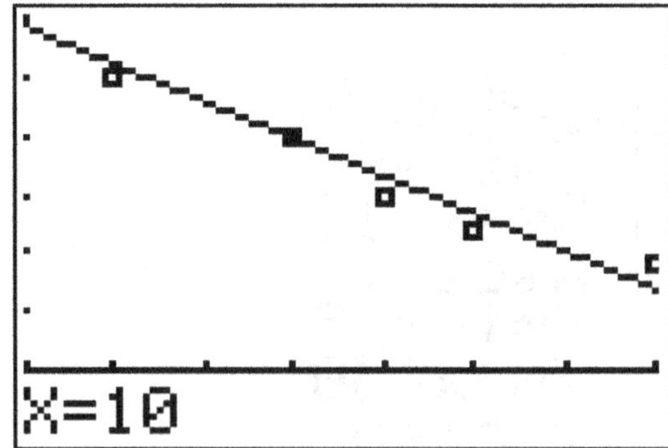

Press ENTER. The predicted value for *y* appears.

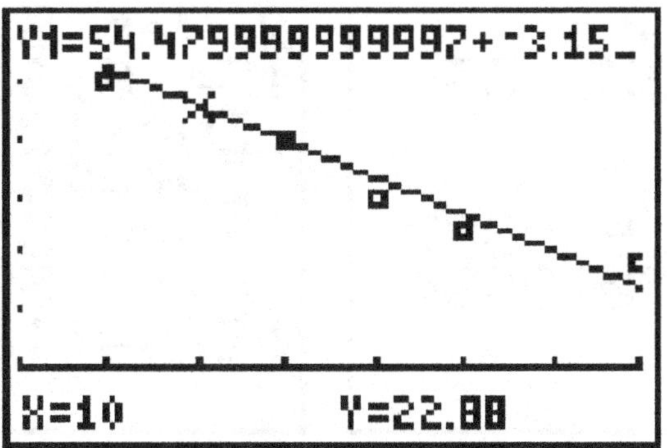

To display the summary statistics for *x* and *y*, press STAT and highlight **CALC**. Select **2:2-Var Stats**. Press ENTER.

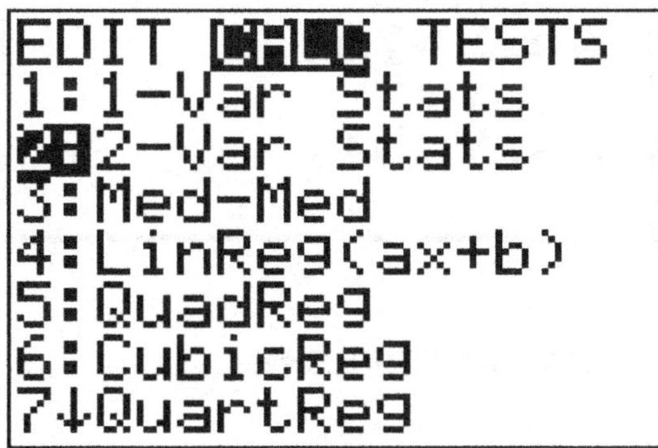

Enter the lists containing the data and press ENTER.

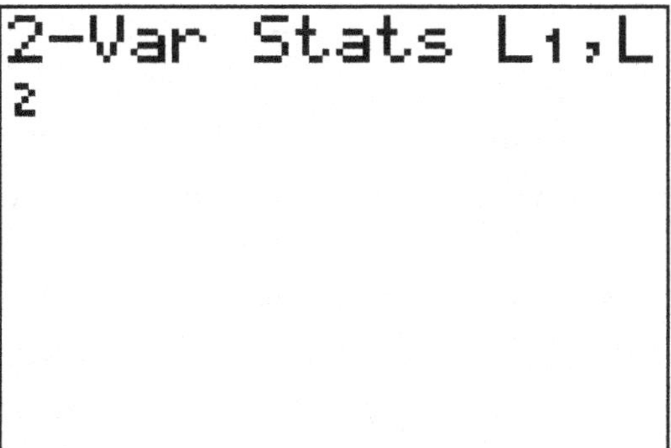

Scroll to the statistics of interest.

```
2-Var Stats
  x̄=11.33333333
  Σx=68
  Σx²=804
  Sx=2.581988897
  σx=2.357022604
↓n=6
```

```
2-Var Stats
↑ȳ=18.66666667
  Σy=112
  Σy²=2436
  Sy=8.310635756
  σy=7.586537784
↓Σxy=1164
```

```
2-Var Stats
↑σy=7.586537784
  Σxy=1164
  minX=8
  maxX=15
  minY=9
  maxY=31
```

LAB ACTIVITIES FOR LINEAR REGRESSION

For each of the following data sets, do the following:

    (a) Enter the data, putting x values in L_1 and y values in L_2.

    (b) Find the equation of the least-squares line.

    (c) Find the value of the correlation coefficient r.

    (d) Draw a scatter diagram and generate the least-squares line.

1. **Cricket Chirps versus Temperature** (data files Slr02L1.txt and Slr02L2.txt)

   In the following data pairs (X, Y), X = chirps/s for the striped ground cricket, and Y = temperature in degrees Fahrenheit.

   Source: *The Song of Insects* by Dr. G. W. Pierce, Harvard College Press

(20.0, 88.6)	(16.0, 71.6)	(19.8, 93.3)	(18.4, 84.3)	(17.1, 80.6)
(15.5, 75.2)	(14.7, 69.7)	(17.1, 82.0)	(15.4, 69.4)	(16.2, 83.3)
(15.0, 79.6)	(17.2, 82.6)	(16.0, 80.6)	(17.0, 83.5)	(14.4, 76.3)

2. **List Price versus Best Price for a New GMC Pickup Truck** (data files Slr01L1.txt and Slr01L2.txt)

In the following data pairs (X, Y), X = list price (in $1000) for a GMC Pickup Truck, and Y = best price (in $1000) for a GMC Pickup Truck.

Source: *Consumer's Digest*, February 1994

(12.400, 11.200)	(14.300, 12.500)	(14.500, 12.700)
(14.900, 13.100)	(16.100, 14.100)	(16.900, 14.800)
(16.500, 14.400)	(15.400, 13.400)	(17.000, 14.900)
(17.900, 15.600)	(18.800, 16.400)	(20.300, 17.700)
(22.400, 19.600)	(19.400, 16.900)	(15.500, 14.000)
(16.700, 14.600)	(17.300, 15.100)	(18.400, 16.100)
(19.200, 16.800)	(17.400, 15.200)	(19.500, 17.000)
(19.700, 17.200)	(21.200, 18.600)	

3. **Diameter of Sand Granules versus Slope on Beach** (data files Slr03L1.txt and Slr03L2.txt)

In the following data pairs (X,Y), X = median diameter (mm) of granules of sand , and Y = gradient of beach slope in degrees. The data are for naturally occurring ocean beaches.

Source: *Physical Geography* by A. M. King, Oxford Press, England

(0.170, 0.630)	(0.190, 0.700)	(0.220, 0.820)
(0.235, 0.880)	(0.235, 1.150)	(0.300, 1.500)
(0.350, 4.400)	(0.420, 7.300)	(0.850, 11.300)

4. **National Unemployment Rate Male versus Female** (data files Slr04L1.txt and Slr04L2.txt)

In the following data pairs (X, Y), X = national unemployment rate for adult males, and Y = national unemployment rate for adult females

Source: *Statistical Abstract of the United States*

(2.9, 4.0)	(6.7, 7.4)	(4.9, 5.0)
(7.9, 7.2)	(9.8, 7.9)	(6.9, 6.1)
(6.1, 6.0)	(6.2, 5.8)	(6.0, 5.2)
(5.1, 4.2)	(4.7, 4.0)	(4.4, 4.4)
(5.8, 5.2)		

CHAPTER 5: ELEMENTARY PROBABILITY THEORY

There are no specific TI-83 Plus or TI-84 Plus activities for the basic rules of probability. However, notice that the TI-83 Plus and TI-84 Plus have menu items for factorial notation, combinations $C_{n,r}$ and permutations $P_{n,r}$. To find these functions, press [MATH] and then highlight **PRB**.

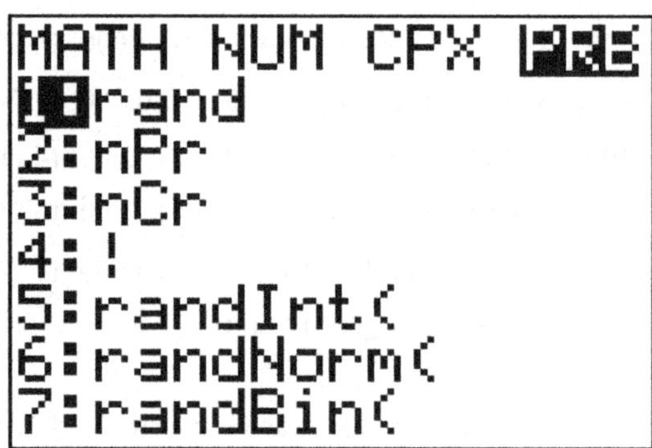

To compute $C_{12,6}$, first clear the screen and type 12. Then use the [MATH] key to select **PRB** and item **3:nCr**. Press [ENTER]. Type the number 6 and then press [ENTER].

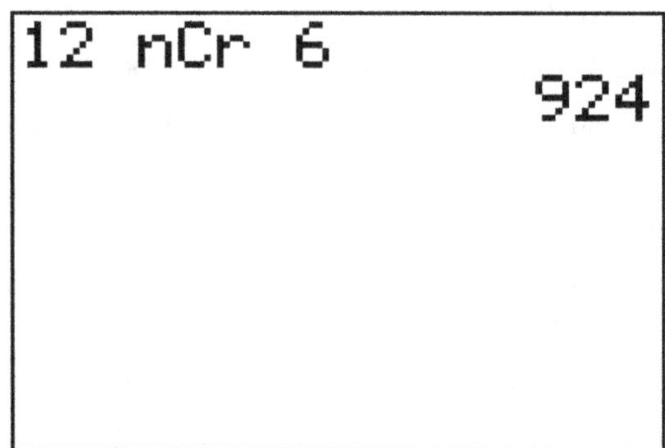

The menu option **2:nPr** works in a similar fashion and gives you the value of $P_{n,r}$.

CHAPTER 6: THE BINOMIAL PROBABILITY DISTRIBUTION AND RELATED TOPICS

DISCRETE PROBABILITY DISTRIBUTIONS (SECTION 6.1 OF *UNDERSTANDING BASIC STATISTICS*)

The TI-83 Plus and TI-84 Plus will compute the mean and standard deviation of a discrete probability distribution. Just enter the values of the random variable x in list L_1 and the corresponding probabilities in list L_2. Use the techniques of grouped data and one-variable statistics to compute the mean and standard deviation.

Example

How long do we spend on hold when we call a store? One study produced the following probability distribution, with the times recorded to the nearest minute.

Time on hold, x:	0	1	2	3	4	5
$P(x)$:	0.15	0.25	0.40	0.10	0.08	0.02

Find the expected value of the time on hold and the standard deviation of the probability distribution.

Enter the times in L_1 and the probabilities in list L_2. Press the [STAT] key, choose **CALC**, and select **1:1-Var Stats** with lists L_1 and L_2. The results are displayed.

```
1-Var Stats
 x̄=1.77
 Σx=1.77
 Σx²=4.53
 Sx=
 σx=1.181989848
↓n=1
```

The expected value is $\bar{x}$ = 1.77 minutes with standard deviation σ = 1.182.

LAB ACTIVITIES FOR DISCRETE PROBABILITY DISTRIBUTIONS

1. The probability distribution for scores on a mechanical aptitude test is as follows:

Score, x:	0	10	20	30	40	50
$P(x)$:	0.130	0.200	0.300	0.170	0.120	0.080

Use the TI-83 Plus or TI-84 Plus to find the expected value and standard deviation of the probability distribution.

2. Hold Insurance has calculated the following probabilities for claims on a new $20,000 vehicle for one
    year if the vehicle is driven by a single male under 25.

$ Claims, x:	0	1000	5000	100,000	200,000
$P(x)$:	0.63	0.24	0.10	0.02	0.01

What is the expected value of the claim in one year? What is the standard deviation of the claim
distribution? What should the annual premium be to include $400 in overhead and profit as well as the
expected claim?

BINOMIAL PROBABILITIES (SECTION 6.2 OF *UNDERSTANDING BASIC STATISTICS*)

For a binomial distribution with n trials, r successes, and the probability of success on a single trial p, the
formula for the probability of r successes out of n trials is

$$P(r) = C_{n,r}\, p^r (1-p)^{n-r}$$

To compute probabilities for specific values of n, r, and p, we can use the TI-83 Plus or TI-84 Plus
graphing calculator.

Example

Consider a binomial experiment with 10 trials and a probability of success on a single trial $p = 0.72$.
Compute the probability of 7 successes out of the 10 trials.

On the TI-83 Plus or TI-84 Plus, the combinations function **nCr** is found in the [MATH] menu under **PRB**.

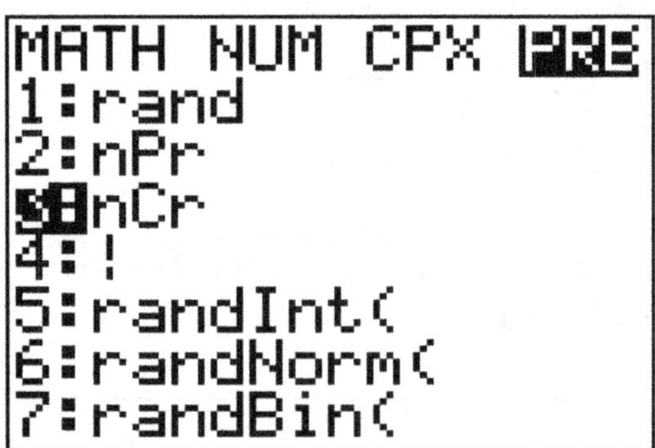

To use the function, we first type the value of n, then **nCr**, and finally the value of r. Here, $n = 10$, $r = 7$,
$p = 0.72$ and $(1-p) = 0.28$. By the formula, we raise 0.72 to the power 7 and 0.28 to the power 3. As shown on
the next page, the probability of 7 successes is 0.264.

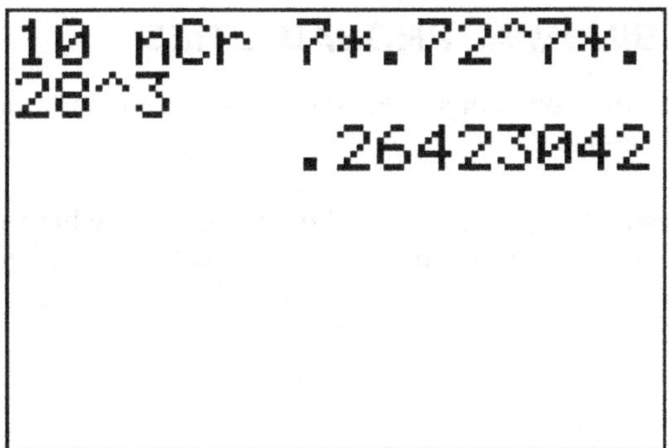

Using the Probability Distributions on the TI-83 Plus and TI-84 Plus

The TI-83 Plus and TI-84 Plus fully support the binomial distribution and have it built in as a function. To access the menu that contains the probability functions, press [2nd] **[DISTR]**. Then scroll to item **A:binompdf(** and press [ENTER].

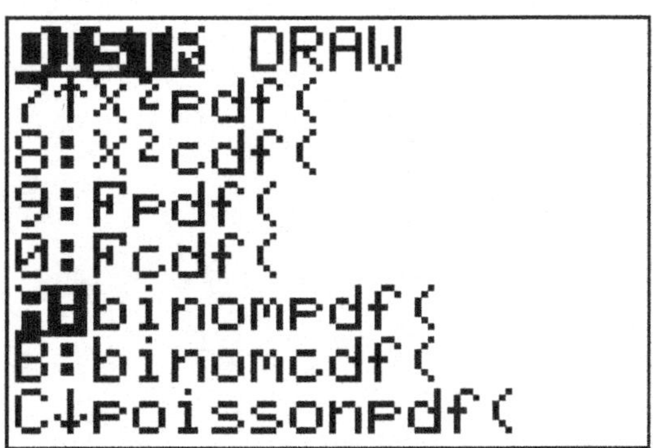

Type in the number of trials, followed by the probability of success on a single trial, followed by the number of successes. Separate each entry by a comma. Press [ENTER].

```
binompdf(10,.72,
7)
            .26423042
```

LAB ACTIVITIES FOR BINOMIAL PROBABILITIES

1. Consider a binomial distribution with $n = 8$ and $p = 0.43$. Use the formula to find the probability of r successes for r from 0 through 8.

2. Consider a binomial distribution with $n = 8$ and $p = 0.43$. Use the built-in binomial probability distribution function to find the probability of r successes for r from 0 through 8.

CHAPTER 7: NORMAL CURVES AND SAMPLING DISTRIBUTIONS

THE AREA UNDER ANY NORMAL CURVE (SECTION 7.3 OF *UNDERSTANDING BASIC STATISTICS*)

The TI-83 Plus and TI-84 Plus give the area under any normal distribution and shade the specified area. Press 2nd [DISTR] to access the probability distributions. Select **2:normalcdf(** and press ENTER.

```
DISTR DRAW
1:normalpdf(
2:normalcdf(
3:invNorm(
4:invT(
5:tpdf(
6:tcdf(
7↓X²pdf(
```

To the right of the function, type in the lower bound, upper bound, μ, and σ in that order, separated by commas, and followed by **)**. To find the area under the normal curve with $\mu = 10$ and $\sigma = 2$ between 4 and 12, select **normalcdf(** and enter the values as shown. Then press ENTER.

```
normalcdf(4,12,1
0,2)
          .8399947732
```

Drawing the Normal Distribution

To draw the graph, first set the window to accommodate the graph. Press the WINDOW button and enter values as shown.

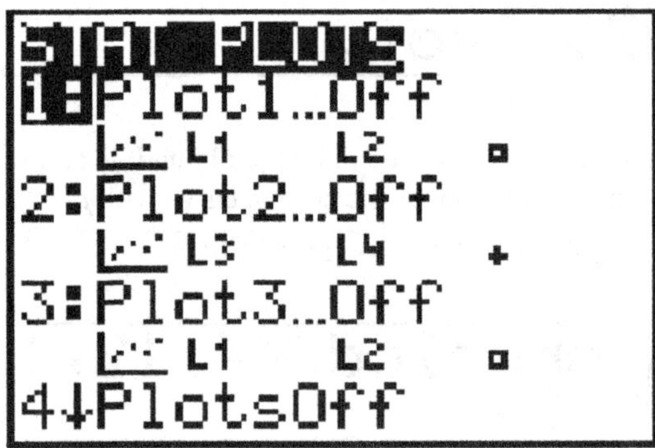

To avoid any stray data points, press 2nd **[STAT PLOT]** and check to be sure all of the plot options are **Off**.

Then press Y= and be sure all entries have been cleared. Press 2nd **[DISTR]** again and highlight **DRAW**. Select **1:ShadeNorm(**.

Again enter the lower limit, upper limit, μ, and σ, separated by commas.

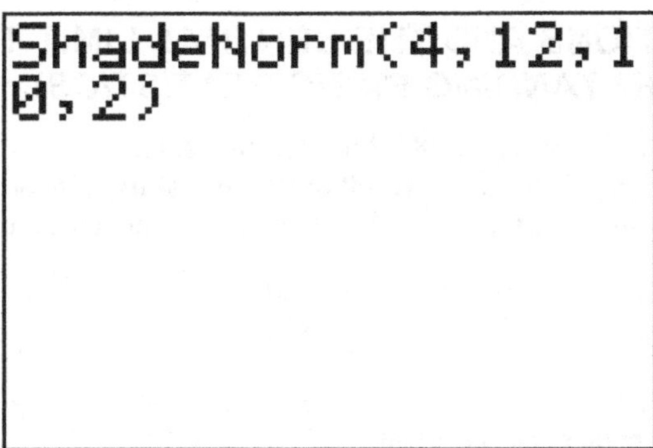

Finally press ENTER.

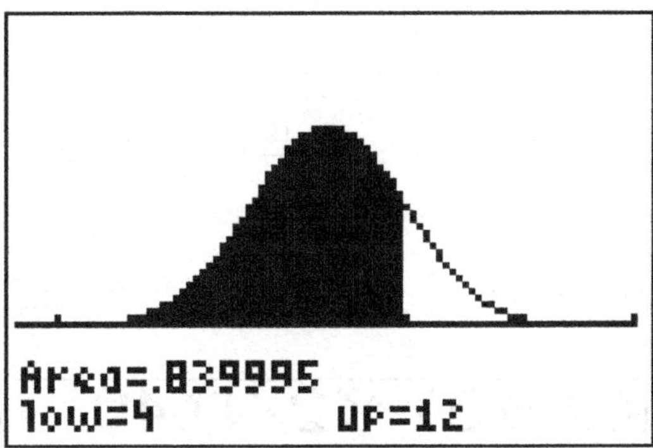

LAB ACTIVITIES FOR THE AREA UNDER ANY NORMAL CURVE

1. Find the area under a normal curve with a mean of 10 and a standard deviation of 2, between 7 and 9. Show the shaded region.

2. To find the area in the right tail of a normal distribution, select the value of 5σ for the upper limit of the region. Find the area that lies to the right of 12 under a normal curve with a mean of 10 and a standard deviation of 2.

3. Consider a random variable x that follows a normal distribution with a mean equal to 100 and a standard deviation equal to 15.

    (a) Shade the region corresponding to $P(x < 90)$ and find the probability.

    (b) Shade the region corresponding to $P(70 < x < 100)$ and find the probability.

    (c) Shade the region corresponding to $P(x > 115)$ and find the probability.

    (d) If the random variable were larger than 145, would that be an unusual event? Explain by computing $P(x > 145)$ and commenting on the meaning of the result.

SAMPLING DISTRIBUTIONS AND THE CENTRAL LIMIT THEOREM (SECTIONS 7.4 AND 7.5 OF *UNDERSTANDING BASIC STATISTICS*)

In this chapter, we use the TI-83 Plus or TI-84 Plus graphing calculator to do statistical computations for normal sampling distributions. For example, as a result of the central limit theorem., we can compute the z score corresponding to a raw score from a distribution of $\bar{x}$ values. The computation uses the formula

$$z = \frac{\bar{x} - \mu}{\frac{\sigma}{\sqrt{n}}}$$

To evaluate z, be sure to use parentheses as necessary.

Example

If a random sample of size 40 is taken from a distribution with mean $\mu = 10$ and standard deviation $\sigma = 2$, find the z score corresponding to $\bar{x} = 9$.

We use the formula

$$z = \frac{9 - 10}{\frac{2}{\sqrt{40}}}$$

Key in the expression using parentheses: $(9 - 10) \div (2 \div \sqrt{}\ (40))$. Press ENTER.

```
(9-10)/(2/√(40))
            -3.16227766
```

The result rounds to $z = -3.16$.

CHAPTER 8: ESTIMATION

CONFIDENCE INTERVALS FOR A POPULATION MEAN (SECTIONS 8.1 AND 8.2 OF *UNDERSTANDING BASIC STATISTICS*)

The TI-83 Plus and TI-84 Plus fully support confidence intervals. To access the confidence interval choices, press STAT and select **TESTS**. The confidence interval choices are found in items 7 through B.

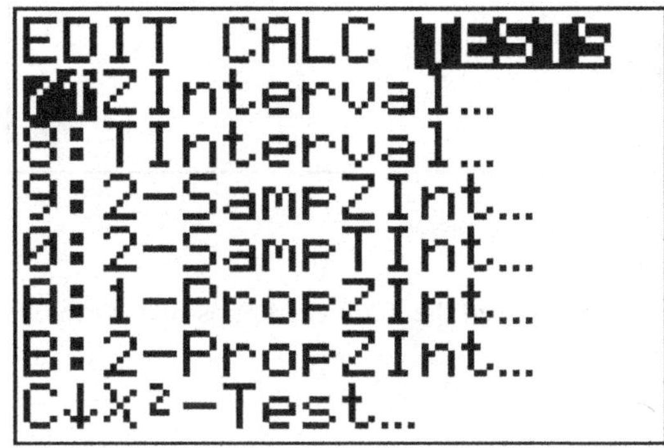

Example (σ is known)

Suppose a random sample of 250 credit card bills showed an average balance of $1200. Also assume that the population standard deviation is $350. Find a 95% confidence interval for the population mean credit card balance.

Because σ is known and we have a large sample, we use the normal distribution. Select **7:ZInterval...**.

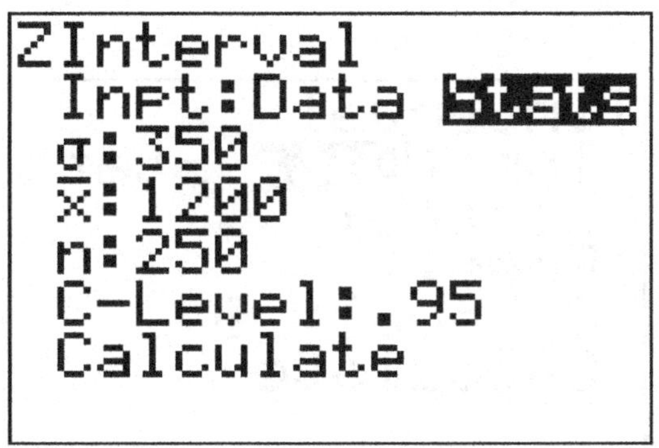

In this example, we have summary statistics, so we will select the **Stats** option for input. We enter the value of σ, the value of $\overline{x}$ and the sample size *n*. Use 0.95 for the C-Level.

Highlight **Calculate** and press ENTER to get the results. Notice that the interval is given using standard mathematical notation for an interval. The interval for μ goes from \$1156.6 to \$1243.4.

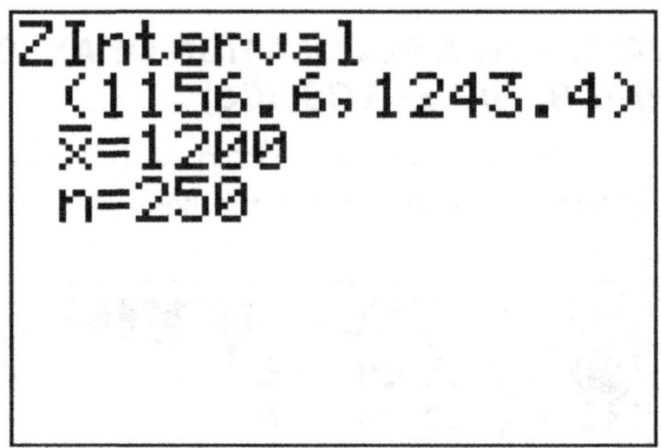

Example (σ is unknown)

A random sample of 16 wolf dens showed the number of pups in each to be as follows:

5 8 7 5 3 4 3 9

5 8 5 6 5 6 4 7

Find a 90% confidence interval for the population mean number of pups in such dens.

In this case we have raw data, so enter the numbers in list L_1 using the **EDIT** option of the STAT key. Because σ is unknown, we use the t distribution. Under **TESTS** from the STAT menu, select item **8:TInterval…**. Because we have raw data, select the **Data** option for Input. The data are in list L_1, and occur with frequency 1. Enter 0.90 for the C-Level.

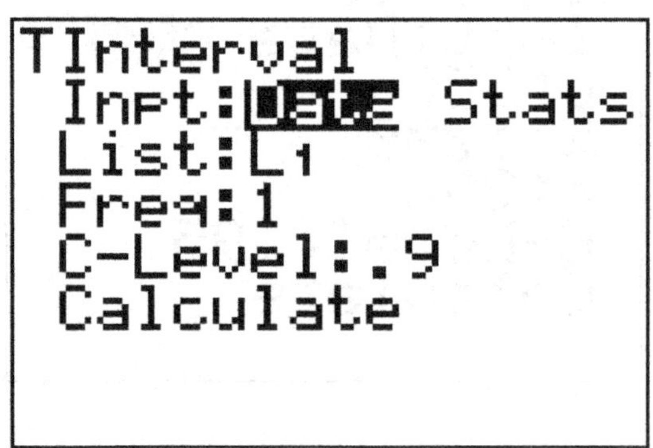

Highlight **Calculate** and press ENTER. The result is the interval from 4.84 pups to 6.41 pups.

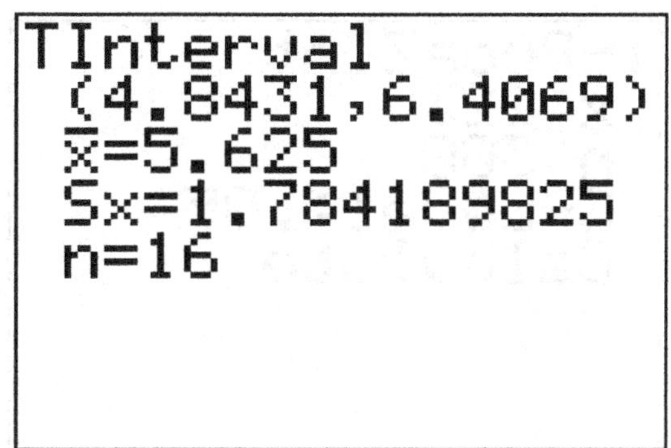

LAB ACTIVITIES FOR CONFIDENCE INTERVALS FOR A POPULATION MEAN

1. Markey Survey was hired to do a study for a new soft drink, Refresh. 20 people were given a can of Refresh and asked to rate it for taste on a scale of 1 to 10 (with 10 being the highest rating). The ratings were

5	8	3	7	5	9	10	6	6	2
9	2	1	8	10	2	5	1	4	7

    Find an 85% confidence interval for the population mean rating of Refresh.

2. Suppose a random sample of 50 basketball players showed the average height to be 78 inches. Also assume that the population standard deviation is 1.5 inches.

    (a) Find a 99% confidence interval for the population mean height.

    (b) Find a 95% confidence interval for the population mean height.

    (c) Find a 90% confidence interval for the population mean height.

    (d) Find an 85% confidence interval for the population mean height.

    (e) What do you notice about the length of the confidence interval as the confidence level goes down? If you used a confidence level of 80%, would you expect the confidence interval to be longer or shorter than that of 85%? Run the program again to verify your answer.

CONFIDENCE INTERVALS FOR THE PROBABILITY OF SUCCESS p IN A BINOMIAL DISTRIBUTION (SECTION 8.3 OF *UNDERSTANDING BASIC STATISTICS*)

To find a confidence interval for a proportion, press the STAT key and use option **A:1-PropZInt...** under **TESTS**. Notice that the normal distribution will be used.

Example

The public television station BPBS wants to find the percent of its viewing population that gives donations to the station. 300 randomly selected viewers were surveyed, and it was found that 123 viewers made contributions to the station. Find a 95% confidence interval for the probability that a viewer of BPBS selected at random contributes to the station.

The letter x is used to count the number of successes (the letter r is used in the text). Enter 123 for x and 300 for n. Use 0.95 for the C-level.

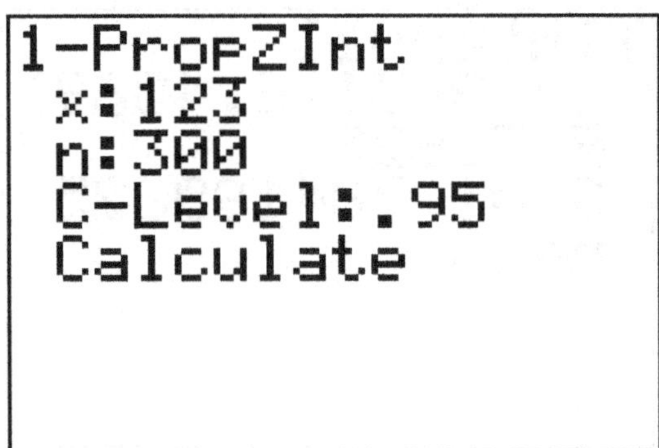

Highlight **Calculate** and press [ENTER]. The result is the interval from 0.35 to 0.47.

LAB ACTIVITIES FOR CONFIDENCE INTERVALS FOR THE PROBABILITY OF SUCCESS *p* IN A BINOMIAL DISTRIBUTION

1. Many types of error will cause a computer program to terminate or give incorrect results. One type of error is punctuation. For instance, if a comma is inserted in the wrong place, the program might not run. A study of programs written by students in an introductory programming course showed that 75 out of 300 errors selected at random were punctuation errors. Find a 99% confidence interval for the proportion of errors made by beginning programming students that are punctuation errors. Next, find a 90% confidence interval. Is this interval longer or shorter?

2. Sam decided to do a statistics project to determine a 90% confidence interval for the probability that a student at West Plains College eats lunch in the school cafeteria. He surveyed a random sample of 12 students and found that 9 ate lunch in the cafeteria. Can Sam use the program to find a confidence interval for the population proportion of students eating in the cafeteria? Why or why not? Try **1-PropZInt** with $n = 12$ and $r = 9$. What happens? What should Sam do to complete his project?

CHAPTER 9: HYPOTHESIS TESTING

The TI-83 Plus and TI-84 Plus both fully support hypothesis testing. Use the $\boxed{\text{STAT}}$ key, then highlight **TESTS**. The options used in Chapter 9 of *Understanding Basic Statistics* are given on the two screens.

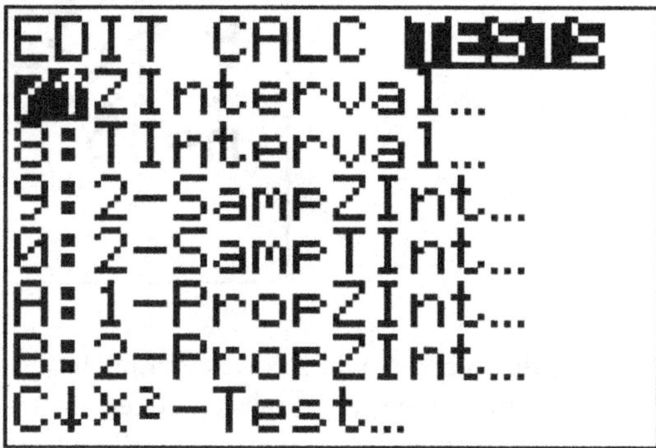

TESTING A SINGLE POPULATION MEAN (SECTIONS 9.1 AND 9.2 OF *UNDERSTANDING BASIC STATISTICS*)

When the value of σ is known, a z-test based on a normal distribution is used to test a population mean, provided that the population has a normal distribution or the sample size is large (at least 30). If the value of σ is unknown, then a t-test will be used if either the population has approximately a normal distribution or the sample size is large. Select option **1:Z-Test...** for a z-test or **2:T-Test...** for a t-test. As with confidence intervals, we have a choice of entering raw data into a list using the **Data** input option or using summary statistics with the **Stats** option. Enter the values for μ_0, $\overline{x}$, σ, and n. The null hypothesis is H_0: $\mu = \mu_0$. To select the alternate hypothesis, use one of the three μ options: $\neq \mu_0$, $< \mu_0$, or $> \mu_0$. The output consists of the value of the sample mean $\overline{x}$, its corresponding test statistic (z) value, and the P-value of the sample statistic.

Example (Testing a mean when σ is known)

Ten years ago, State College did a study regarding the number of hours full-time students worked each week. The mean number of hours was 8.7. A recent study involving a random sample of 45 full-time students showed that the average number of hours worked per week was 10.3. Assume that the population standard deviation was 2.8. Use a 5% level of confidence to test if the mean number of hours worked per week by full-time students has increased.

We have a large sample, so we will use the normal distribution for our sample test statistic. Select option **1:Z-Test...**. We have summary statistics, so use the **Stats** input option. Enter the values as shown below.

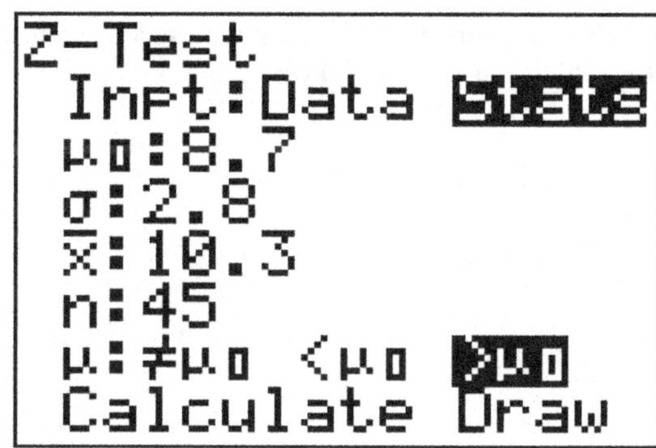

Next highlight **Calculate** and press $\boxed{\text{ENTER}}$.

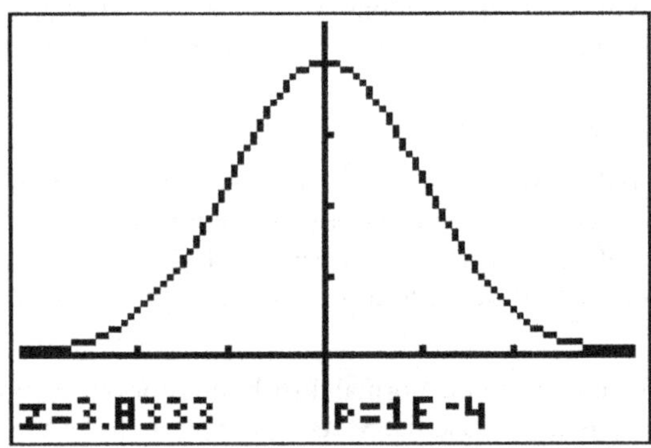

We see that the value for z corresponding to the sample mean $\bar{x} = 10.3$ is $z = 3.83$. The critical value for a 5% level of significance and a right-tailed test is $z_0 = 1.645$. Clearly, the sample test statistic z is to the right of z_0. Therefore, we reject H_0. Notice that the P-value = 6.325E-5. This means that we move the decimal 5 places to the left, giving a P-value of 0.000063. Because the P-value is less than 0.05, we reject H_0.

Graph

The TI-83 Plus and TI-84 Plus both provide an option to show the sample test statistic on the normal distribution. Highlight the **Draw** option on the **Z-Test** screen. To avoid any stray data points, press $\boxed{\text{2nd}}$ **[STAT PLOT]** and check to be sure all of the plot options are **Off**. Then press $\boxed{\text{Y=}}$ and be sure all entries have been cleared.

Because z is so far to the right, it does not appear on this window. However, its value does, and a rounded P-value appears as well.

To do hypothesis testing of the mean when σ is unknown, we use a Student's t distribution. Select option **2:T-Test...**. The entry screens are similar to those for **Z-Test**.

LAB ACTIVITIES FOR TESTING A SINGLE POPULATION MEAN

1. A random sample of 65 pro basketball players showed their heights (in feet) to be as follows:

6.50	6.25	6.33	6.50	6.42	6.67	6.83	6.82
6.17	7.00	5.67	6.50	6.75	6.54	6.42	6.58
6.00	6.75	7.00	6.58	6.29	7.00	6.92	6.42
5.92	6.08	7.00	6.17	6.92	7.00	5.92	6.42
6.00	6.25	6.75	6.17	6.75	6.58	6.58	6.46
5.92	6.58	6.13	6.50	6.58	6.63	6.75	6.25
6.67	6.17	6.17	6.25	6.00	6.75	6.17	6.83
6.00	6.42	6.92	6.50	6.33	6.92	6.67	6.33
6.08							

   (a) Enter the data into list L_1. Use **1-Var Stats** to determine the sample standard deviation.

   (b) Use the **T-Test** option to test the hypothesis that the average height of the players is greater than 6.2 feet, at the 1% level of significance.

2. In this problem, we will see how the test conclusion is possibly affected by a change in the level of significance.

   Teachers for Excellence is interested in the attention span of students in grades 1 and 2, now as compared to 20 years ago. They believe it has decreased. Studies done 20 years ago indicate that the attention span of children in grades 1 and 2 was 15 minutes. A study sponsored by Teachers for Excellence involved a random sample of 20 students in grades 1 and 2. The average attention span of these students was (in minutes) $\bar{x} = 14.2$ with standard deviation $s = 1.5$.

   (a) Conduct the hypothesis test using $\alpha = 0.05$ and a left-tailed test. What is the test conclusion? What is the P-value?

   (b) Conduct the hypothesis test using $\alpha = 0.01$ and a left-tailed test. What is the test conclusion? How could you have predicted this result by looking at the P-value from part (a)? Is the P-value for this part the same as it was for part (a)?

3. In this problem, let's explore the effect that sample size has on the process of testing a mean. Run **Z-Test** with the hypotheses H_0: $\mu_0 = 200$, H_1: $\mu > 200$, $\alpha = 0.05$, $\bar{x} = 210$, and $\sigma = 40$.

   (a) Use the sample size $n = 30$. Note the P-value, z score of the sample test statistic, and test conclusion.

   (b) Use the sample size $n = 50$. Note the P-value and z score of the sample test statistic, and test conclusion.

   (c) Use the sample size $n = 100$. Note the P-value and z score of the sample test statistic, and test conclusion.

   (d) In general, if your sample statistic is close to the proposed population mean specified in H_0 and you want to reject H_0, would you use a smaller or a larger sample size?

TESTS INVOLVING A SINGLE PROPORTION (SECTION 9.3 OF *UNDERSTANDING BASIC STATISTICS*)

To conduct a hypotheses test of a single proportion, select option **5:1-PropZTest...**. The null hypothesis is H_0: $p = p_0$ (the letter k is used in the text instead of p_0). Enter the value of p_0. The number of successes is designated by the value x (r in the text). Enter the value of x. The sample size, or number of trials, is n. The alternate hypothesis will be one of the three proportion (prop) options: $\neq p_0$, $< p_0$, or $> p_0$. Highlight the appropriate choice. Finally highlight **Calculate** and press ENTER. Notice that the **Draw** option is available to show the results on the standard normal distribution.

CHAPTER 10: INFERENCES ABOUT DIFFERENCES

TESTS INVOLVING PAIRED DIFFERENCES (DEPENDENT SAMPLES) (SECTION 10.1 OF *UNDERSTANDING BASIC STATISTICS*)

To perform a paired difference test, we put our paired data into two columns, and then put the differences between corresponding pairs of values in a third column. For example, put the "before" data in list L_1 and the "after" data in L_2. Create $L_3 = L_1 - L_2$.

Example

Promoters of a state lottery decided to advertise the lottery heavily on television for one week during the middle of one of the lottery games. To see if the advertising improved ticket sales, the promoters surveyed a random sample of 8 ticket outlets and recorded weekly sales for one week before the television campaign and for one week after the campaign. The results (in ticket sales) follow, where B stands for "before" and A for "after" the advertising campaign.

B:	3201	4529	1425	1272	1784	1733	2563	3129
A:	3762	4851	1202	1131	2172	1802	2492	3151

At the 0.05 level of significance, test the claim that the television campaign increased lottery-ticket sales.

We want to test to see if $\mathbf{D = B - A}$ is less than zero, as we are testing the claim that the lottery ticket sales are greater after the television campaign. We will put the "before" data in L_1 and the "after" data in L_2.

```
L1          L2          ■          3
3201        3762        ------
4529        4851
1425        1202
1272        1131
1784        2172
1733        1802
2563        2492

L3 =L1−L2
```

The differences in list L_3 can be generated by highlighting the column header **L3**, entering $L_1 - L_2$, and pressing [ENTER].

```
L1          L2          ▮▮      3
3201        3762        -561
4529        4851        -322
1425        1202        223
1272        1131        141
1784        2172        -388
1733        1802        -69
2563        2492        71
L3 ={-561,-322,2...
```

Next, we will conduct a *t*-test on the differences in list L_3. Select option **2:T-Test...** from the **TESTS** menu. Select **Data** for the **Inpt** option. The null hypotheses states that the difference $d = 0$, so we use the null hypothesis H_0: $d = \mu_0$, with $\mu_0 = 0$. Select L_3 for **List.**, and, because the test is a left-tailed test, select $\mu < \mu_0$.

```
T-Test
 Inpt:Data Stats
 µo:0
 List:L3
 Freq:1
 µ:≠µo  <µo  >µo
Calculate  Draw
```

Highlight **Calculate** and press ENTER.

```
T-Test
 µ<0
 t=-1.178455137
 P=.138559057
 x̄=-115.875
 Sx=278.1132542
 n=8
```

We see that the sample *t* value is –1.178 with a corresponding *P*-value of 0.139. Because the *P*-value is greater than 0.05, we do not reject H_0.

LAB ACTIVITIES USING TESTS INVOLVING PAIRED DIFFERENCES (DEPENDENT SAMPLES)

1. The data are pairs of values where the first entry represents average salary (in thousands of dollars/year) for male faculty members at an institution and the second entry represents the average salary for female faculty members (in thousands of dollars/year) at the same institution. A random sample of 22 U.S. colleges and universities was used (source: *Academe, Bulletin of the American Association of University Professors*).

$$
\begin{array}{lllll}
(34.5, 33.9) & (30.5, 31.2) & (35.1, 35.0) & (35.7, 34.2) & (31.5, 32.4) \\
(34.4, 34.1) & (32.1, 32.7) & (30.7, 29.9) & (33.7, 31.2) & (35.3, 35.5) \\
(30.7, 30.2) & (34.2, 34.8) & (39.6, 38.7) & (30.5, 30.0) & (33.8, 33.8) \\
(31.7, 32.4) & (32.8, 31.7) & (38.5, 38.9) & (40.5, 41.2) & (25.3, 25.5) \\
(28.6, 28.0) & (35.8, 35.1)
\end{array}
$$

   (a) Put the first entries in L_1, the second in L_2, and create L_3 to be the difference $L_1 - L_2$.

   (b) Use the **T-Test** option to test the hypothesis that there is a difference in salary. What is the *P*-value of the sample test statistic? Do we reject or fail to reject the null hypothesis at the 5% level of significance? What about at the 1% level of significance?

   (c) Use the **T-Test** option to test the hypothesis that female faculty members have a lower average salary than male faculty members. What is the test conclusion at the 5% level of significance? At the 1% level of significance?

2. An audiologist is conducting a study on noise and stress. Twelve subjects selected at random were given a stress test in a room that was quiet. Then the same subjects were given another stress test, this time in a room with high-pitched background noise. The results of the stress tests were scores 1 through 20, with 20 indicating the greatest stress. The results, where B represents the score of the test administered in the quiet room and A represents the scores of the test administered in the room with the high-pitched background noise, are shown below.

Subject	1	2	4	5	6	7	8	9	10	11	12
B	13	12	16	19	7	13	9	15	17	6	14
A	18	15	14	18	10	12	11	14	17	8	16

   Test the hypothesis that the stress level was greater during exposure to noise. Look at the *P*-value. Should you reject the null hypotheses at the 1% level of significance? At the 5% level?

INFERENCES ABOUT THE DIFFERENCE OF TWO MEANS $\mu_1 - \mu_2$ AND THE DIFFERENCE OF TWO PROPORTIONS $p_1 - p_2$ (SECTIONS 10.2 AND 10.3 OF *UNDERSTANDING BASIC STATISTICS*)

Tests of difference of means for independent samples are presented in Section 10.2 of *Understanding Basic Statistics*. We consider the $\bar{x}_1 - \bar{x}_2$ distribution. The null hypothesis is that there is no difference between means, so $H_0: \mu_1 = \mu_2$, or $H_0: \mu_1 - \mu_2 = 0$.

Testing the Difference of Two Means When σ_1 and σ_2 are Known

In general, for tests of difference of means when σ_1 and σ_2 are known, use the **2-SampZTest** from the **TESTS** menu. For the difference of means, you have the option of using either summary statistics or raw data in lists as input. **Data** allows you to use raw data from a list, while **Stats** lets you use summary statistics.

One of the tests for independent samples allows you to find confidence intervals for $\mu_1 - \mu_2$. Press the STAT key, highlight **TESTS** and use the option **9:2-SampZInt...** for confidence intervals when σ_1 and σ_2 are known. Note that, if the populations are not normal distributions, large samples are required.

Example

A random sample of 45 pro football players produced a sample mean height of 6.18 feet with population standard deviation 0.37 feet. A random sample of 40 pro basketball players produced a mean height of 6.45 feet with population standard deviation 0.31. Find a 90% confidence interval for the difference of population heights.

Because σ_1 and σ_2 are known and we have large samples (both greater than 30), we can use the normal distribution. Therefore, we select option **9:2-SampZInt...**. Because we have summary statistics, we select **Stats** for the **Inpt** option. Then we enter the specified values. The data entry requires two screens, as shown below.

```
2-SampZInt          2-SampZInt
 Inpt:Data Stats    ↑σ2:.31
 σ1:.37              x̄1:6.18
 σ2:.31              n1:45
 x̄1:6.18             x̄2:6.45
 n1:45               n2:40
 x̄2:6.45             C-Level:.9
↓n2:40               Calculate
```

Highlight **Calculate** and press ENTER.

```
2-SampZInt
 (-.3914, -.1486)
 x̄1=6.18
 x̄2=6.45
 n1=45
 n2=40
```

The interval is from -0.39 to -0.15. Because zero lies outside the interval, we conclude that at the 90% level, the mean height of basketball players is different from that of football players.

Testing the Difference of Two Means When σ_1 and σ_2 are Unknown

In general, for tests of difference of means when σ_1 and σ_2 are unknown, use the **2-SampTTest** from the **TESTS** menu. For the difference of means, you have the option of using either summary statistics or raw data in lists as input. **Data** allows you to use raw data from a list, while **Stats** lets you use summary statistics.

For confidence intervals for $\mu_1 - \mu_2$ when σ_1 and σ_2 are unknown, use the option **2-SampTInt** from the **TESTS** menu. Use sample estimates s_1 and s_2 for corresponding population standard deviations σ_1 and σ_2. When working with small samples, be sure to choose **Yes** for the option **Pooled**.

Example

Sellers of microwave French fry cookers claim that their process saves cooking time. McDougle Fast Food Chain is considering the purchase of these new cookers, but wants to test the claim. Six batches of French fries were cooked in the traditional way. These times (in minutes) are as follows:

15   17   14   15   16   13

Six batches of French fries of the same weight were cooked using the new microwave cooker. These cooking times (in minutes) are as follows:

11   14   12   10   11   15

Test the claim that the microwave process takes less time. Use $\alpha = 0.05$.

Put the data for traditional cooking in list L_1 and the data for the new method in List L_2. Because we have small samples, we will want to use the Student's t distribution. Select option **4:2-SampTTest...**. Select **Data** for the **Inpt** option, choose the alternate hypothesis $\mu_1 > \mu_2$ and select **Yes** for **Pooled**.

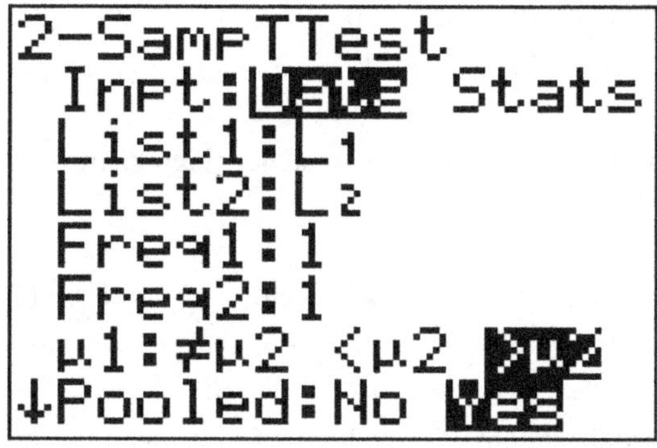

Highlight **Calculate** and press [ENTER]. The output is on two screens, as shown on the next page.

```
2-SampTTest        2-SampTTest
 µ1>µ2              µ1>µ2
 t=2.890086704     ↑Sx1=1.41421356
 P=.0080523896      Sx2=1.94079022
 df=10              SxP=1.69803808
 x̄1=15              n1=6
↓x̄2=12.16666667     n2=6
```

Because the *P*-value is 0.008, which is less than 0.05, we reject H_0 and conclude that the new method cooks food faster.

Testing the difference of two proportions ($p_1 - p_2$)

Tests of differences of proportions are presented in Section 10.3 of *Understanding Basic Statistics*. To conduct a hypothesis test for a difference of proportions, use the **2-PropZTest** from the **TESTS** menu. Option **B:2-PropZInt...** gives confidence intervals for the difference of two proportions.

LAB ACTIVITIES FOR INFERENCES ABOUT THE DIFFERENCE OF MEANS (INDEPENDENT SAMPLES) AND THE DIFFERENCE OF PROPORTIONS

1. Calm Cough Medicine is testing a new ingredient to see if its addition will lengthen the effective cough relief time of a single dose. A random sample of 15 doses of the standard medicine was tested, and the effective relief times were (in minutes):

   42   35   40   32   30   26   51   39   33   28
   37   22   36   33   41

   A random sample of 20 doses was tested when the new ingredient was added. The effective relief times were (in minutes):

   43   51   35   49   32   29   42   38   45   74
   31   31   46   36   33   45   30   32   41   25

   Assume that the standard deviations of the relief times are equal for the two populations. Test the claim that the effective relief time is longer when the new ingredient is added. Use $\alpha = 0.01$.

2. Publisher's Survey did a study to see if the proportion of men who read mysteries is different from the proportion of women who read them. A random sample of 402 women showed that 112 read mysteries regularly (at least six books per year). A random sample of 365 men showed that 92 read mysteries regularly. Is the proportion of mystery readers different between men and women? Use a 1% level of significance.

   (a) Find the *P*-value of the test conclusion.

   (b) Test the hypothesis that the proportion of women who read mysteries is *greater* than the proportion of men. Use a 1% level of significance. Is the *P*-value for a right-tailed test half that of a two-tailed test? If you know the *P*-value for a two-tailed test, can you draw conclusions for a one-tailed test?

3. The following data are based on random samples of red foxes in two regions of Germany. The number of cases of rabies was counted for a random sample of 16 areas in each of two regions.

Region 1:	10	2	2	5	3	4	3	3
Region 2:	1	1	2	1	3	9	2	2

Region 1:	4	0	2	6	4	8	7	4
Region 2:	3	4	3	2	2	0	0	2

(a) Use a confidence level $c = 90\%$. Does the interval indicate that the population mean number of rabies cases in Region 1 is greater than the population mean number of rabies cases in Region 2 at the 90% level? Why or why not?

(b) Use a confidence level $c = 95\%$. Does the interval indicate that the population mean number of rabies cases in Region 1 is greater than the population mean number of rabies cases in Region 2 at the 95% level? Why or why not?

(c) Determine whether the population mean number of rabies cases in Region 1 is greater than the population mean number of rabies cases in Region 2 at the 99% level. Explain your answer. Verify your answer by running **2-SampTInt** with a 99% confidence level.

4. A random sample of 30 police officers working the night shift showed that 23 used at least 5 sick leave days per year. Another random sample of 45 police officers working the day shift showed that 26 used at least 5 sick leave days per year. Find the 90% confidence interval for the difference of population proportions of police officers working the two shifts and using at least 5 sick leave days per year. At the 90% level, does it seem that the proportions are different? Does there seem to be a difference in proportions at the 99% level? Why or why not?

CHAPTER 11: ADDITIONAL TOPICS USING INFERENCE

CHI-SQUARE TEST OF INDEPENDENCE (SECTION 11.1 OF *UNDERSTANDING BASIC STATISTICS*)

The TI-83 Plus and TI-84 Plus calculators support tests for independence. Press the [STAT] key, select **TESTS**, and scroll down to option **C:χ^2-Test...**. You will then need to enter the original observed values into a matrix.

Example

A computer programming aptitude test has been developed for high school seniors. The test designers claim that scores on the test are independent of the type of school the student attends: rural, suburban, or urban. A study involving a random sample of students from each of these types of institutions yielded the following information, where aptitude scores range from 200 to 500 with 500 indicating the greatest aptitude and 200 the least. The entry in each cell is the observed number of students achieving the indicated score on the test.

School Type

Score	Rural	Suburban	Urban
200–299	33	65	82
300–399	45	79	95
400–500	21	47	63

Using the option **C:χ^2-Test...**, test the claim that the aptitude test scores are independent of the type of school attended at the 0.05 level of significance.

To use a matrix to enter the data, press [2nd] [MATRIX]. Highlight **EDIT**, and select **1:[A]**. Press [ENTER].

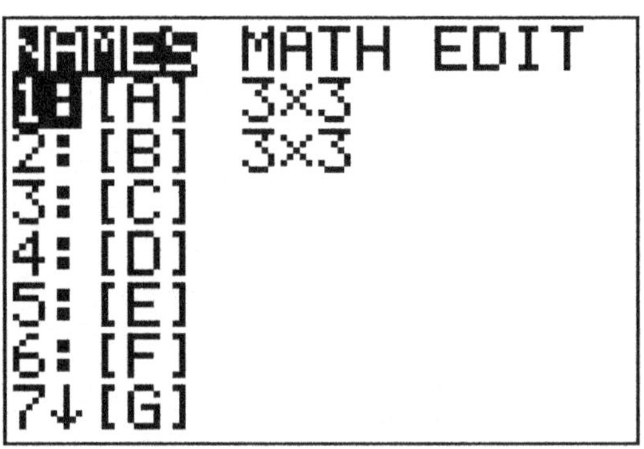

Because the contingency table has 3 rows and 3 columns, type the number 3 (for number of rows), press [ENTER], type 3 again (for number of columns), and press [ENTER]. Then type in the observed values. Press [ENTER] after each entry. Notice that you enter the table by rows. When the table is entered completely, press [2nd] **[MATRIX]**   again.

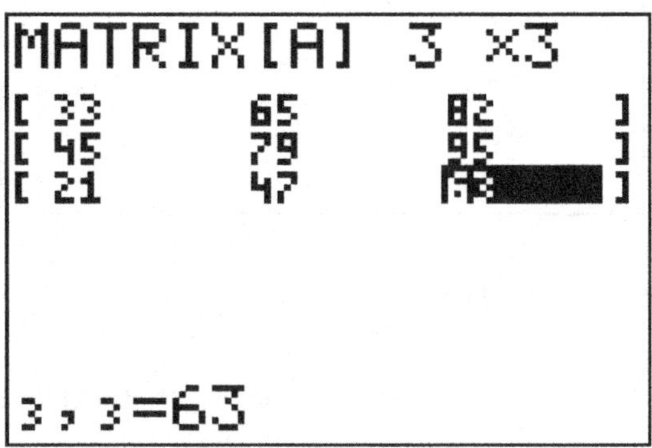

This time, select **2:[B]** from the **EDIT** menu. Set the dimensions to be the same as matrix [A]. For this example, [B] should have 3 rows and 3 columns.

Now press the [STAT] key, highlight **TESTS**, and use option **C:χ^2-Test....** Press [ENTER]. The information on the screen tells you that the observed values of the table are in matrix [A], and that the expected values will be placed in matrix [B].

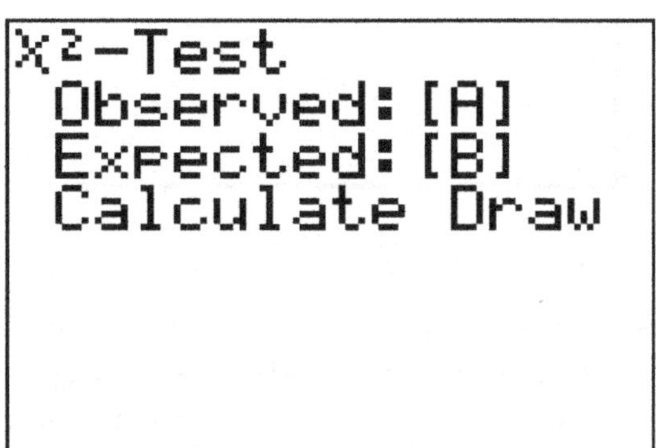

To change the designation of either matrix, say from [B] to [C], delete the matrix name ([B]), press [2nd] **[MATRIX]**, in the **NAMES** menu select the name of the matrix you want (**3:[C]**), and press [ENTER].

Once you have the right matrix designations for the Observed and Expected values, highlight **Calculate** and press [ENTER].

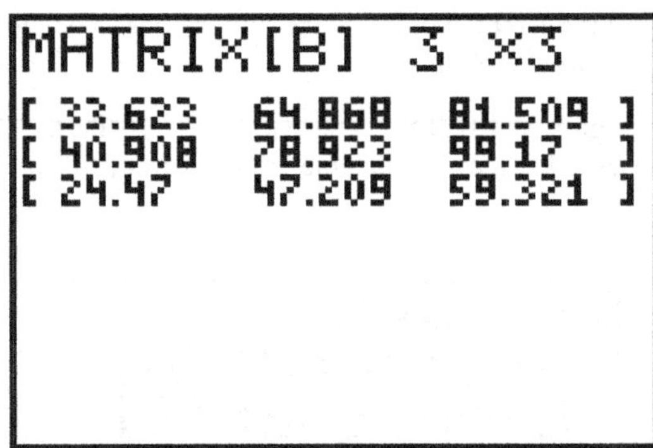

We see that the sample value of χ^2 is 1.32. The *P*-value is 0.858. Because the *P*-value is larger than 0.05, we do not reject the null hypothesis.

If you want to see the expected values in matrix [B], type 2nd [MATRIX], and in the **EDIT** menu select **2:[B]**.

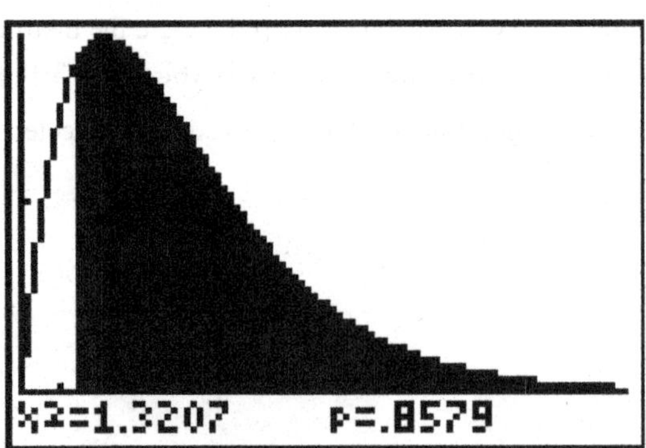

Graph

Notice that one of the options of the χ^2 test is to graph the χ^2 distribution and show the sample test statistic on the graph. Highlight the **Draw** option on the χ^2-**Test** screen, and press ENTER. Before you do this, press 2nd [STAT PLOT] to be sure that all plot options are set to **Off**. Also press Y= and be sure all entries have been cleared.

LAB ACTIVITIES FOR CHI-SQUARE TEST OF INDEPENDENCE

1. We Care Auto Insurance had its staff of actuaries conduct a study to see if vehicle type and loss claim are independent. A random sample of auto claims over six months gives the information in the contingency table.

Total Loss Claims per Year per Vehicle

Type of vehicle	$0–999	$1000–2999	$3000–5999	$6000+
Sports car	20	10	16	18
Truck	16	25	33	9
Family Sedan	40	68	17	7
Compact	52	73	48	12

Test the claim that car type and loss claim are independent. Use $\alpha = 0.05$.

2. An educational specialist is interested in comparing three methods of instruction:

SL – standard lecture with discussion

TV – video taped lectures with no discussion

IM – individualized method with reading assignments and tutoring, but no lectures

The specialist conducted a study of these three methods to see if they are independent. A course was taught using each of the three methods and a standard final exam was given at the end. Students were put into the different method sections at random. The course type and test results are shown in the following contingency table.

Final Exam Score

Course Type	< 60	60–69	70–79	80–89	90–100
SL	10	4	70	31	25
TV	8	3	62	27	23
IM	7	2	58	25	22

Test the claim that the instruction method and final test scores are independent, using $\alpha = 0.01$.

Appendix: Descriptions of Data Sets in the Online Study Center

Appendix Preface

There are over 70 data sets saved in Excel, Minitab Portable, SPSS, TI-83 Plus, and TI-84 Plus/ASCII formats to accompany *Understanding Basic Statistics,* 5th edition. These files can be found on the Brase/Brase statistics site at www.cengage.com/statistics/Brase/UBS5e. The data sets are organized by category.

    A.   The following are provided for each data set:
        1.   The category
        2.   A brief description of the data and variables with a reference when appropriate
        3.   File names for Excel, Minitab, SPSS, and TI-83 Plus and TI-84 Plus/ASCII formats
    B.   The categories are
        1.   **Single variable**
            File name prefix **Sv** followed by the data set number
            41 data sets..page A-5
        2.   **Two variable independent samples** (large and small samples)
            File name prefix **Tvis** followed by the data set number
            10 data sets..page A-19
        3.   **Two variable dependent samples** appropriate for *t*-tests
            File name prefix **Tvds** followed by the data set number
            10 data sets..page A-24
        4.   **Simple linear regression**
            File name prefix **Slr** followed by the data set number
            12 data sets..page A-28
    C.   The formats are
        1.   Excel files in subdirectory Excel_5e. These files have suffix .xls
        2.   Minitab portable files in subdirectory Minitab_5e. These files have suffix .mtp
        3.   TI-83 Plus and TI-84 Plus/ASCII files in subdirectory TI8384_5e. These files have suffix .txt
        4.   SPSS files in subdirectory SPSS_5e. These files have suffix .sav

        A-3

Suggestions for Using the Data Sets

1. Single variable (file name prefix Sv)

   These data sets are appropriate for:

   Graphs: Histograms, box plots

   Descriptive statistics: Mean, median, mode, variance, standard deviation, coefficient of variation, 5-number summary

   Inferential statistics: Confidence intervals for the population mean, hypothesis tests of a single mean

2. Two independent data sets (file name prefix Tvis)

   Graphs: Histograms, box plots for each data set

   Descriptive statistics: Mean, median, mode, variance, standard deviation, coefficient of variation, 5-number summary for each data set

   Inferential statistics: Confidence intervals for the difference of means, hypothesis tests for the difference of means

3. Paired data, dependent samples (file name prefix Tvds)

   Descriptive statistics: Mean, median, mode, variance, standard deviation, coefficient of variation, 5-number summary for the difference of the paired data values.

   Inferential statistics: Hypothesis tests for the difference of means (paired data)

4. Data pairs for simple linear regression (file name prefix Slr)

   Graphs: Scatter plots, for individual variables histograms and box plots

   Descriptive statistics:

   - Mean, median, mode, variance, standard deviation, coefficient of variation, 5-number summary for individual variables.

   - Least-squares line, sample correlation coefficient, sample coefficient of determination

Descriptions of Data Sets

SINGLE VARIABLE

File name prefix: Sv followed by the number of the data file

01. Disney Stock Volume (Single Variable)
The following data represent the number of shares of Disney stock (in hundreds of shares) sold for a random sample of 60 trading days
Reference: *The Denver Post*, business section

12584	9441	18960	21480	10766	13059	8589	4965
4803	7240	10906	8561	6389	14372	18149	6309
13051	12754	10860	9574	19110	29585	21122	14522
17330	18119	10902	29158	16065	10376	10999	17950
15418	12618	16561	8022	9567	9045	8172	13708
11259	10518	9301	5197	11259	10518	9301	5197
6758	7304	7628	14265	13054	15336	14682	27804
16022	24009	32613	19111				

File names	Excel: Sv01.xls
	Minitab: Sv01.mtp
	SPSS: Sv01.sav
	TI-83 Plus and TI-84 Plus/ASCII: Sv01.txt

02. Weights of Pro Football Players (Single Variable)
The following data represent weights in pounds of 50 randomly selected pro football linebackers.
Reference: The Sports Encyclopedia of Pro Football

225	230	235	238	232	227	244	222
250	226	242	253	251	225	229	247
239	223	233	222	243	237	230	240
255	230	245	240	235	252	245	231
235	234	248	242	238	240	240	240
235	244	247	250	236	246	243	255
241	245						

File names	Excel: Sv02.xls
	Minitab: Sv02.mtp
	SPSS: Sv02.sav
	TI-83 Plus and TI-84 Plus/ASCII: Sv02.txt

03. Heights of Pro Basketball Players (Single Variable)
The following data represent heights in feet of 65 randomly selected pro basketball players.
Reference: All-Time Player Directory, The Official NBA Encyclopedia

6.50	6.25	6.33	6.50	6.42	6.67	6.83	6.82
6.17	7.00	5.67	6.50	6.75	6.54	6.42	6.58
6.00	6.75	7.00	6.58	6.29	7.00	6.92	6.42
5.92	6.08	7.00	6.17	6.92	7.00	5.92	6.42
6.00	6.25	6.75	6.17	6.75	6.58	6.58	6.46
5.92	6.58	6.13	6.50	6.58	6.63	6.75	6.25
6.67	6.17	6.17	6.25	6.00	6.75	6.17	6.83
6.00	6.42	6.92	6.50	6.33	6.92	6.67	6.33
6.08							

File names   Excel: Sv03.xls
     Minitab: Sv03.mtp
     SPSS: Sv03.sav
     TI-83 Plus and TI-84 Plus/ASCII: Sv03.txt

04. Miles per Gallon Gasoline Consumption (Single Variable)
The following data represent miles per gallon gasoline consumption (highway) for a random sample
of 55 makes and models of passenger cars.
Reference: Environmental Protection Agency

30	27	22	25	24	25	24	15
35	35	33	52	49	10	27	18
20	23	24	25	30	24	24	24
18	20	25	27	24	32	29	27
24	27	26	25	24	28	33	30
13	13	21	28	37	35	32	33
29	31	28	28	25	29	31	

File names   Excel: Sv04.xls
     Minitab: Sv04.mtp
     SPSS: Sv04.sav
     TI-83 Plus and TI-84 Plus/ASCII: Sv04.txt

05. Fasting Glucose Blood Tests (Single Variable)
The following data represent glucose blood levels (mg/100mL) after a 12-hour fast for a random
sample of 70 women.
Reference: *American J. Clin. Nutr.*, Vol. 19, pp. 345-351

45	66	83	71	76	64	59	59
76	82	80	81	85	77	82	90
87	72	79	69	83	71	87	69
81	76	96	83	67	94	101	94
89	94	73	99	93	85	83	80
78	80	85	83	84	74	81	70
65	89	70	80	84	77	65	46
80	70	75	45	101	71	109	73
73	80	72	81	63	74		

File names   Excel: Sv05.xls
     Minitab: Sv05.mtp
     SPSS: Sv05.sav
     TI-83 Plus and TI-84 Plus/ASCII: Sv05.txt

06. Number of Children in Rural Canadian Families (Single Variable)
The following data represent the numbers of children in a random sample of 50 rural Canadian families.
Reference: *American Journal of Sociology*, Vol. 53, pp. 470-480

11	13	4	14	10	2	5	0
0	3	9	2	5	2	3	3
3	4	7	1	9	4	3	3
2	6	0	2	6	5	9	5
4	3	2	5	2	2	3	5
14	7	6	6	2	5	3	4
6	1						

File names	Excel: Sv06.xls
	Minitab: Sv06.mtp
	SPSS: Sv06.sav
	TI-83 Plus and TI-84 Plus/ASCII: Sv06.txt

07. Children as a % of Population (Single Variable)
The following data represent percentages of children in the population for a random sample of 72 Denver neighborhoods.
Reference: The Piton Foundation, Denver, Colorado

30.2	18.6	13.6	36.9	32.8	19.4	12.3	39.7	22.2	31.2
36.4	37.7	38.8	28.1	18.3	22.4	26.5	20.4	37.6	23.8
22.1	53.2	6.8	20.7	31.7	10.4	21.3	19.6	41.5	29.8
14.7	12.3	17.0	16.7	20.7	34.8	7.5	19.0	27.2	16.3
24.3	39.8	31.1	34.3	15.9	24.2	20.3	31.2	30.0	33.1
29.1	39.0	36.0	31.8	32.9	26.5	4.9	19.5	21.0	24.2
12.1	38.3	39.3	20.2	24.0	28.6	27.1	30.0	60.8	39.2
21.6	20.3								

File names	Excel: Sv07.xls
	Minitab: Sv07.mtp
	SPSS: Sv07.sav
	TI-83 Plus and TI-84 Plus/ASCII: Sv07.txt

08. Percentage Change in Household Income (Single Variable)
The following data represent the percentage change in household income over a five-year period for a random sample of $n = 78$ Denver neighborhoods.
Reference: The Piton Foundation, Denver, Colorado

27.2	25.2	25.7	80.9	26.9	20.2	25.4	26.9	26.4	26.3
27.5	38.2	20.9	31.3	23.5	26.0	35.8	30.9	15.5	24.8
29.4	11.7	32.6	32.2	27.6	27.5	28.7	28.0	15.6	20.0
21.8	18.4	27.3	13.4	14.7	21.6	26.8	20.9	32.7	29.3
21.4	29.0	7.2	25.7	25.5	39.8	26.6	24.2	33.5	16.0
29.4	26.8	32.0	24.7	24.2	29.8	25.8	18.2	26.0	26.2
21.7	27.0	23.7	28.0	11.2	26.2	21.6	23.7	28.3	34.1
40.8	16.0	50.5	54.1	3.3	23.5	10.1	14.8		

File names	Excel: Sv08.xls
	Minitab: Sv08.mtp
	SPSS: Sv08.sav
	TI-83 Plus and TI-84 Plus/ASCII: Sv08.txt

09. Crime Rate per 1,000 Population (Single Variable)

The following data represent the crime rates per 1,000 population for a random sample of 70 Denver neighborhoods.

Reference: The Piton Foundation, Denver, Colorado

84.9	45.1	132.1	104.7	258.0	36.3	26.2	207.7
58.5	65.3	42.5	53.2	172.6	69.2	179.9	65.1
32.0	38.3	185.9	42.4	63.0	86.4	160.4	26.9
154.2	111.0	139.9	68.2	127.0	54.0	42.1	105.2
77.1	278.0	73.0	32.1	92.7	704.1	781.8	52.2
65.0	38.6	22.5	157.3	63.1	289.1	52.7	108.7
66.3	69.9	108.7	96.9	27.1	105.1	56.2	80.1
59.6	77.5	68.9	35.2	65.4	123.2	130.8	70.7
25.1	62.6	68.6	334.5	44.6	87.1		

File names	Excel: Sv09.xls
	Minitab: Sv09.mtp
	SPSS: Sv09.sav
	TI-83 Plus and TI-84 Plus/ASCII: Sv09.txt

10. Percentage Change in Population (Single Variable)

The following data represent the percentage change in population over a nine-year period for a random sample of 64 Denver neighborhoods.

Reference: The Piton Foundation, Denver, Colorado

6.2	5.4	8.5	1.2	5.6	28.9	6.3	10.5	-1.5	17.3
21.6	-2.0	-1.0	3.3	2.8	3.3	28.5	-0.7	8.1	32.6
68.6	56.0	19.8	7.0	38.3	41.2	4.9	7.8	7.8	97.8
5.5	21.6	32.5	-0.5	2.8	4.9	8.7	-1.3	4.0	32.2
2.0	6.4	7.1	8.8	3.0	5.1	-1.9	-2.6	1.6	7.4
10.8	4.8	1.4	19.2	2.7	71.4	2.5	6.2	2.3	10.2
1.9	2.3	-3.3	2.6						

File names	Excel: Sv10.xls
	Minitab: Sv10.mtp
	SPSS: Sv10.sav
	TI-83 Plus and TI-84 Plus/ASCII: Sv10.txt

11. Thickness of the Ozone Column (Single Variable)

The following data represent the January mean thickness of the ozone column above Arosa, Switzerland (Dobson units: one milli-centimeter ozone at standard temperature and pressure). The data is from a random sample of years from 1926 on.

Reference: Laboratorium für Atmosphärensphysik, Switzerland

324	332	362	383	335	349	354	319	360	329
400	341	315	368	361	336	349	347	338	332
341	352	342	361	318	337	300	352	340	371
327	357	320	377	338	361	301	331	334	387
336	378	369	332	344					

File names	Excel: Sv11.xls
	Minitab: Sv11.mtp
	SPSS: Sv11.sav
	TI-83 Plus and TI-84 Plus/ASCII: Sv11.txt

12. **Sun Spots (Single Variable)**
The following data represent the January mean number of sunspots. The data are taken from a random sample of Januarys from 1749 to 1983.
Reference: Waldmeir, M., *Sun Spot Activity*, International Astronomical Union Bulletin

12.5	14.1	37.6	48.3	67.3	70.0	43.8	56.5	59.7	24.0
12.0	27.4	53.5	73.9	104.0	54.6	4.4	177.3	70.1	54.0
28.0	13.0	6.5	134.7	114.0	72.7	81.2	24.1	20.4	13.3
9.4	25.7	47.8	50.0	45.3	61.0	39.0	12.0	7.2	11.3
22.2	26.3	34.9	21.5	12.8	17.7	34.6	43.0	52.2	47.5
30.9	11.3	4.9	88.6	188.0	35.6	50.5	12.4	3.7	18.5
115.5	108.5	119.1	101.6	59.9	40.7	26.5	23.1	73.6	165.0
202.5	217.4	57.9	38.7	15.3	8.1	16.4	84.3	51.9	58.0
74.7	96.0	48.1	51.1	31.5	11.8	4.5	78.1	81.6	68.9

File names   Excel: Sv12.xls
Minitab: Sv12.mtp
SPSS: Sv12.sav
TI-83 Plus and TI-84 Plus/ASCII: Sv12.txt

13. **Motion of Stars (Single Variable)**
The following data represent the angular motions of stars across the sky due to each star's own velocity. A random sample of stars from the M92 globular cluster was used. Units are arc seconds per century.
Reference: Cudworth, K.M., *Astronomical Journal*, Vol. 81, pp. 975-982

0.042	0.048	0.019	0.025	0.028	0.041	0.030	0.051	0.026
0.040	0.018	0.022	0.048	0.045	0.019	0.028	0.029	0.018
0.033	0.035	0.019	0.046	0.021	0.026	0.026	0.033	0.046
0.023	0.036	0.024	0.014	0.012	0.037	0.034	0.032	0.035
0.015	0.027	0.017	0.035	0.021	0.016	0.036	0.029	0.031
0.016	0.024	0.015	0.019	0.037	0.016	0.024	0.029	0.025
0.022	0.028	0.023	0.021	0.020	0.020	0.016	0.016	0.016
0.040	0.029	0.025	0.025	0.042	0.022	0.037	0.024	0.046
0.016	0.024	0.028	0.027	0.060	0.045	0.037	0.027	0.028
0.022	0.048	0.053						

File names   Excel: Sv13.xls
Minitab: Sv13.mtp
SPSS: Sv13.sav
TI-83 Plus and TI-84 Plus/ASCII: Sv13.txt

14. **Arsenic and Ground Water (Single Variable)**
    The following data represent (naturally occurring) concentrations of arsenic in ground water for a random sample of 102 Northwest Texas wells. Units are parts per billion.
    Reference: Nichols, C.E. and Kane, V.E., Union Carbide Technical Report K/UR-1

7.6	10.4	13.5	4.0	19.9	16.0	12.0	12.2	11.4	12.7
3.0	10.3	21.4	19.4	9.0	6.5	10.1	8.7	9.7	6.4
9.7	63.0	15.5	10.7	18.2	7.5	6.1	6.7	6.9	0.8
73.5	12.0	28.0	12.6	9.4	6.2	15.3	7.3	10.7	15.9
5.8	1.0	8.6	1.3	13.7	2.8	2.4	1.4	2.9	13.1
15.3	9.2	11.7	4.5	1.0	1.2	0.8	1.0	2.4	4.4
2.2	2.9	3.6	2.5	1.8	5.9	2.8	1.7	4.6	5.4
3.0	3.1	1.3	2.6	1.4	2.3	1.0	5.4	1.8	2.6
3.4	1.4	10.7	18.2	7.7	6.5	12.2	10.1	6.4	10.7
6.1	0.8	12.0	28.1	9.4	6.2	7.3	9.7	62.1	15.5
6.4	9.5								

File names	Excel: Sv14.xls
	Minitab: Sv14.mtp
	SPSS: Sv14.sav
	TI-83 Plus and TI-84 Plus/ASCII: Sv14.txt

15. **Uranium in Ground Water (Single Variable)**
    The following data represent (naturally occurring) concentrations of uranium in ground water for a random sample of 100 Northwest Texas wells. Units are parts per billion.
    Reference: Nichols, C.E. and Kane, V.E., Union Carbide Technical Report K/UR-1

8.0	13.7	4.9	3.1	78.0	9.7	6.9	21.7	26.8
56.2	25.3	4.4	29.8	22.3	9.5	13.5	47.8	29.8
13.4	21.0	26.7	52.5	6.5	15.8	21.2	13.2	12.3
5.7	11.1	16.1	11.4	18.0	15.5	35.3	9.5	2.1
10.4	5.3	11.2	0.9	7.8	6.7	21.9	20.3	16.7
2.9	124.2	58.3	83.4	8.9	18.1	11.9	6.7	9.8
15.1	70.4	21.3	58.2	25.0	5.5	14.0	6.0	11.9
15.3	7.0	13.6	16.4	35.9	19.4	19.8	6.3	2.3
1.9	6.0	1.5	4.1	34.0	17.6	18.6	8.0	7.9
56.9	53.7	8.3	33.5	38.2	2.8	4.2	18.7	12.7
3.8	8.8	2.3	7.2	9.8	7.7	27.4	7.9	11.1
24.7								

File names	Excel: Sv15.xls
	Minitab: Sv15.mtp
	SPSS: Sv15.sav
	TI-83 Plus and TI-84 Plus/ASCII: Sv15.txt

16.   Ground Water pH (Single Variable)
A pH less than 7 is acidic, and a pH above 7 is alkaline. The following data represent pH levels in
ground water for a random sample of 102 Northwest Texas wells.
Reference: Nichols, C.E. and Kane, V.E., Union Carbide Technical Report K/UR-1

```
7.6   7.7   7.4   7.7   7.1   8.2   7.4   7.5   7.2   7.4
7.2   7.6   7.4   7.8   8.1   7.5   7.1   8.1   7.3   8.2
7.6   7.0   7.3   7.4   7.8   8.1   7.3   8.0   7.2   8.5
7.1   8.2   8.1   7.9   7.2   7.1   7.0   7.5   7.2   7.3
8.6   7.7   7.5   7.8   7.6   7.1   7.8   7.3   8.4   7.5
7.1   7.4   7.2   7.4   7.3   7.7   7.0   7.3   7.6   7.2
8.1   8.2   7.4   7.6   7.3   7.1   7.0   7.0   7.4   7.2
8.2   8.1   7.9   8.1   8.2   7.7   7.5   7.3   7.9   8.8
7.1   7.5   7.9   7.5   7.6   7.7   8.2   8.7   7.9   7.0
8.8   7.1   7.2   7.3   7.6   7.1   7.0   7.0   7.3   7.2
7.8   7.6
```

File names	Excel: Sv16.xls
	Minitab: Sv16.mtp
	SPSS: Sv16.sav
	TI-83 Plus and TI-84 Plus/ASCII: Sv16.txt

17.   Static Fatigue 90% Stress Level (Single Variable)
Kevlar Epoxy is a material used on the NASA space shuttle. Strands of this epoxy were tested at 90%
breaking strength. The following data represent time to failure in hours at the 90% stress level for a
random sample of 50 epoxy strands.
Reference: R.E. Barlow, University of California, Berkeley

```
0.54   1.80   1.52   2.05   1.03   1.18   0.80   1.33   1.29   1.11
3.34   1.54   0.08   0.12   0.60   0.72   0.92   1.05   1.43   3.03
1.81   2.17   0.63   0.56   0.03   0.09   0.18   0.34   1.51   1.45
1.52   0.19   1.55   0.02   0.07   0.65   0.40   0.24   1.51   1.45
1.60   1.80   4.69   0.08   7.89   1.58   1.64   0.03   0.23   0.72
```

File names	Excel: Sv17.xls
	Minitab: Sv17.mtp
	SPSS: Sv17.sav
	TI-83 Plus and TI-84 Plus/ASCII: Sv17.txt

18.   Static Fatigue 80% Stress Level (Single Variable)
Kevlar Epoxy is a material used on the NASA space shuttle. Strands of this epoxy were tested at 80%
breaking strength. The following data represent time to failure in hours at the 80% stress level for a
random sample of 54 epoxy strands.
Reference: R.E. Barlow, University of California, Berkeley

```
152.2   166.9   183.8     8.5     1.8   118.0   125.4   132.8    10.6
 29.6    50.1   202.6   177.7   160.0    87.1   112.6   122.3   124.4
131.6   140.9     7.5    41.9    59.7    80.5    83.5   149.2   137.0
301.1   329.8   461.5   739.7   304.3   894.7   220.2   251.0   269.2
130.4    77.8    64.4   381.3   329.8   451.3   346.2   663.0    49.1
 31.7   116.8   140.2   334.1   285.9    59.7    44.1   351.2    93.2
```

File names	Excel: Sv18.xls
	Minitab: Sv18.mtp
	SPSS: Sv18.sav
	TI-83 Plus and TI-84 Plus/ASCII: Sv18.txt

19. Tumor Recurrence (Single Variable)
Certain kinds of tumors tend to recur. The following data represent the lengths of time in months for a tumor to recur after chemotherapy (sample size: 42).
Reference: Byar, D.P., *Urology,* Vol. 10, pp. 556-561

19	18	17	1	21	22	54	46	25	49
50	1	59	39	43	39	5	9	38	18
14	45	54	59	46	50	29	12	19	36
38	40	43	41	10	50	41	25	19	39
27	20								

File names   Excel: Sv19.xls
Minitab: Sv19.mtp
SPSS: Sv19.sav
TI-83 Plus and TI-84 Plus/ASCII: Sv19.txt

20. Weight of Harvest (Single Variable)
The following data represent the weights in kilograms of maize harvest from a random sample of 72 experimental plots on the island of St. Vincent (Caribbean).
Reference: Springer, B.G.F., *Proceedings, Caribbean Food Corps. Soc.,* Vol. 10, pp. 147-152

24.0	27.1	26.5	13.5	19.0	26.1	23.8	22.5	20.0
23.1	23.8	24.1	21.4	26.7	22.5	22.8	25.2	20.9
23.1	24.9	26.4	12.2	21.8	19.3	18.2	14.4	22.4
16.0	17.2	20.3	23.8	24.5	13.7	11.1	20.5	19.1
20.2	24.1	10.5	13.7	16.0	7.8	12.2	12.5	14.0
22.0	16.5	23.8	13.1	11.5	9.5	22.8	21.1	22.0
11.8	16.1	10.0	9.1	15.2	14.5	10.2	11.7	14.6
15.5	23.7	25.1	29.5	24.5	23.2	25.5	19.8	17.8

File names   Excel: Sv20.xls
Minitab: Sv20.mtp
SPSS: Sv20.sav
TI-83 Plus and TI-84 Plus/ASCII: Sv20.txt

21. Apple Trees (Single Variable)
The following data represent the trunk girths (mm) from a random sample of 60 four-year-old apple trees at East Malling Research Station (England)
Reference: S.C. Pearce, University of Kent at Canterbury

108	99	106	102	115	120	120	117	122	142
106	111	119	109	125	108	116	105	117	123
103	114	101	99	112	120	108	91	115	109
114	105	99	122	106	113	114	75	96	124
91	102	108	110	83	90	69	117	84	142
122	113	105	112	117	122	129	100	138	117

File names   Excel: Sv21.xls
Minitab: Sv21.mtp
SPSS: Sv21.sav
TI-83 Plus and TI-84 Plus/ASCII: Sv21.txt

22. **Black Mesa Archaeology (Single Variable)**
    The following data represent rim diameters (cm) from a random sample of 40 bowls found at Black
    Mesa archaeological site. The diameters are estimated from broken pot shards.
    Reference: Michelle Hegmon, Crow Canyon Archaeological Center, Cortez, Colorado

```
17.2  15.1  13.8  18.3  17.5  11.1   7.3  23.1  21.5  19.7
17.6  15.9  16.3  25.7  27.2  33.0  10.9  23.8  24.7  18.6
16.9  18.8  19.2  14.6   8.2   9.7  11.8  13.3  14.7  15.8
17.4  17.1  21.3  15.2  16.8  17.0  17.9  18.3  14.9  17.7
```

File names   Excel: Sv22.xls
                        Minitab: Sv22.mtp
                        SPSS: Sv22.sav
                        TI-83 Plus and TI-84 Plus/ASCII: Sv22.txt

23. **Wind Mountain Archaeology (Single Variable)**
    The following data represent depths (cm) for a random sample of 73 significant archaeological
    artifacts at the Wind Mountain excavation site.
    Reference: Woosley, A. and McIntyre, A., *Mimbres Mogolion Archaeology*, University of New
    Mexico Press

```
85    45   75    60    90    90   115    30    55    58
78   120   80    65    65   140    65    50    30   125
75   137   80   120    15    45    70    65    50    45
95    70   70    28    40   125   105    75    80    70
90    68   73    75    55    70    95    65   200    75
15    90   46    33   100    65    60    55    85    50
10    68   99   145    45    75    45    95    85    65
65    52   82
```

File names   Excel: Sv23.xls
                        Minitab: Sv23.mtp
                        SPSS: Sv23.sav
                        TI-83 Plus and TI-84 Plus/ASCII: Sv23.txt

24. **Arrow Heads (Single Variable)**
    The following data represent the lengths (cm) of a random sample of 61 projectile points found at the
    Wind Mountain Archaeological site.
    Reference: Woosley, A. and McIntyre, A., *Mimbres Mogolion Archaeology*, University of New
    Mexico Press

```
3.1  4.1  1.8  2.1  2.2  1.3  1.7  3.0  3.7  2.3
2.6  2.2  2.8  3.0  3.2  3.3  2.4  2.8  2.8  2.9
2.9  2.2  2.4  2.1  3.4  3.1  1.6  3.1  3.5  2.3
3.1  2.7  2.1  2.0  4.8  1.9  3.9  2.0  5.2  2.2
2.6  1.9  4.0  3.0  3.4  4.2  2.4  3.5  3.1  3.7
3.7  2.9  2.6  3.6  3.9  3.5  1.9  4.0  4.0  4.6
1.9
```

File names   Excel: Sv24.xls
                        Minitab: Sv24.mtp
                        SPSS: Sv24.sav
                        TI-83 Plus and TI-84 Plus/ASCII: Sv24.txt

25. **Anasazi Indian Bracelets (Single Variable)**
    The following data represent the diameters (cm) of shell bracelets and rings found at the Wind Mountain archaeological site.
    Reference: Woosley, A. and McIntyre, A., *Mimbres Mogolion Archaeology*, University of New Mexico Press

```
5.0   5.0   8.0   6.1   6.0   5.1   5.9   6.8   4.3   5.5
7.2   7.0   5.0   5.6   5.3   7.0   3.4   8.2   4.3   5.2
1.5   6.1   4.0   6.0   5.5   5.2   5.2   5.2   5.5   7.2
6.0   6.2   5.2   5.0   4.0   5.7   5.1   6.1   5.7   7.3
7.3   6.7   4.2   4.0   6.0   7.1   7.3   5.5   5.8   8.9
7.5   8.3   6.8   4.9   4.0   6.2   7.7   5.0   5.2   6.8
6.1   7.2   4.4   4.0   5.0   6.0   6.2   7.2   5.8   6.8
7.7   4.7   5.3
```

File names	Excel: Sv25.xls
	Minitab: Sv25.mtp
	SPSS: Sv25.sav
	TI-83 Plus and TI-84 Plus/ASCII: Sv25.txt

26. **Pizza Franchise Fees (Single Variable)**
    The following data represent annual franchise fees (in thousands of dollars) for a random sample of 36 pizza franchises.
    Reference: *Business Opportunities Handbook*

```
25.0   15.5    7.5   19.9   18.5   25.5   15.0    5.5   15.2   15.0
14.9   18.5   14.5   29.0   22.5   10.0   25.0   35.5   22.1   89.0
17.5   33.3   17.5   12.0   15.5   25.5   12.5   17.5   12.5   35.0
30.0   21.0   35.5   10.5    5.5   20.0
```

File names	Excel: Sv26.xls
	Minitab: Sv26.mtp
	SPSS: Sv26.sav
	TI-83 Plus and TI-84 Plus/ASCII: Sv26.txt

27. **Pizza Franchise Start-up Requirement (Single Variable)**
    The following data represent the start-up costs (in thousands of dollars) for a random sample of 36 pizza franchises.
    Reference: *Business Opportunities Handbook*

```
40    25    50   129   250   128   110   142    25    90
75   100   500   214   275    50   128   250    50    75
30    40   185    50   175   125   200   150   150   120
95    30   400   149   235   100
```

File names	Excel: Sv27.xls
	Minitab: Sv27.mtp
	SPSS: Sv27.sav
	TI-83 Plus and TI-84 Plus/ASCII: Sv27.txt

28. **College Degrees (Single Variable)**
     The following data represent percentages of the adult population with college degrees. The sample is from a random collection of 68 Midwest counties.
     Reference: *County and City Data Book,* 12th edition, U.S. Department of Commerce

```
 9.9    9.8    6.8    8.9   11.2   15.5    9.8   16.8    9.9   11.6
 9.2    8.4   11.3   11.5   15.2   10.8   16.3   17.0   12.8   11.0
 6.0   16.0   12.1    9.8    9.4    9.9   10.5   11.8   10.3   11.1
12.5    7.8   10.7    9.6   11.6    8.8   12.3   12.2   12.4   10.0
10.0   18.1    8.8   17.3   11.3   14.5   11.0   12.3    9.1   12.7
 5.6   11.7   16.9   13.7   12.5    9.0   12.7   11.3   19.5   30.7
 9.4    9.8   15.1   12.8   12.9   17.5   12.3    8.2
```

File names	Excel: Sv28.xls
	Minitab: Sv28.mtp
	SPSS: Sv28.sav
	TI-83 Plus and TI-84 Plus/ASCII: Sv28.txt

29. **Poverty Level (Single Variable)**
     The following data represent percentages of all persons below the poverty level. The sample is from a random collection of 80 cities in the Western U.S.
     Reference: *County and City Data Book,* 12th edition, U.S. Department of Commerce

```
12.1   27.3   20.9   14.9    4.4   21.8    7.1   16.4   13.1
 9.4    9.8   15.7   29.9    8.8   32.7    5.1    9.0   16.8
21.6    4.2   11.1   14.1   30.6   15.4   20.7   37.3    7.7
19.4   18.5   19.5    8.0    7.0   20.2    6.3   12.9   13.3
30.0    4.9   14.4   14.1   22.6   18.9   16.8   11.5   19.2
21.0   11.4    7.8    6.0   37.3   44.5   37.1   28.7    9.0
17.9   16.0   20.2   11.5   10.5   17.0    3.4    3.3   15.6
16.6   29.6   14.9   23.9   13.6    7.8   14.5   19.6   31.5
28.1   19.2    4.9   12.7   15.1    9.6   23.8   10.1
```

File names	Excel: Sv29.xls
	Minitab: Sv29.mtp
	SPSS: Sv29.sav
	TI-83 Plus and TI-84 Plus/ASCII: Sv29.txt

30. **Working at Home (Single Variable)**
     The following data represent percentages of adults whose primary employment involves working at home. The data are from a random sample of 50 California cities.
     Reference: *County and City Data Book,* 12th edition, U.S. Department of Commerce

```
4.3    5.1    3.1    8.7    4.0    5.2   11.8    3.4    8.5    3.0
4.3    6.0    3.7    3.7    4.0    3.3    2.8    2.8    2.6    4.4
7.0    8.0    3.7    3.3    3.7    4.9    3.0    4.2    5.4    6.6
2.4    2.5    3.5    3.3    5.5    9.6    2.7    5.0    4.8    4.1
3.8    4.8   14.3    9.2    3.8    3.6    6.5    2.6    3.5    8.6
```

File names	Excel: Sv30.xls
	Minitab: Sv30.mtp
	SPSS: Sv30.sav
	TI-83 Plus and TI-84 Plus/ASCII: Sv30.txt

31. **Number of Pups in Wolf Den (Single Variable)**
    The following data represent numbers of wolf pups per den from a random sample of 16 wolf dens.
    Reference: *The Wolf in the Southwest: The Making of an Endangered Species*, Brown, D.E.,
    University of Arizona Press

5	8	7	5	3	4	3	9
5	8	5	6	5	6	4	7

    File names   Excel: Sv31.xls
                      Minitab: Sv31.mtp
                      SPSS: Sv31.sav
                      TI-83 Plus and TI-84 Plus/ASCII: Sv31.txt

32. **Glucose Blood Level (Single Variable)**
    The following data represent glucose blood levels (mg/100ml) after a 12-hour fast for a random
    sample of 6 tests given to an individual adult female.
    Reference: *American J. Clin. Nutr., Vol. 19*, pp. 345-351

83	83	86	86	78	88

    File names   Excel: Sv32.xls
                      Minitab: Sv32.mtp
                      SPSS: Sv32.sav
                      TI-83 Plus and TI-84 Plus/ASCII: Sv32.txt

33. **Length of Remission (Single Variable)**
    The drug 6-mP (6-mercaptopurine) is used to treat leukemia. The following data represent the lengths
    of remission in weeks for a random sample of 21 patients using 6-mP.
    Reference: E.A. Gehan, University of Texas Cancer Center

10	7	32	23	22	6	16	34	32	25
11	20	19	6	17	35	6	13	9	6
10									

    File names   Excel: Sv33.xls
                      Minitab: Sv33.mtp
                      SPSS: Sv33.sav
                      TI-83 Plus and TI-84 Plus/ASCII: Sv33.txt

34. **Entry Level Jobs (Single Variable)**
    The following data represent percentages of entry-level jobs in a random sample of 16 Denver
    neighborhoods.
    Reference: The Piton Foundation, Denver, Colorado

8.9	22.6	18.5	9.2	8.2	24.3	15.3	3.7
9.2	14.9	4.7	11.6	16.5	11.6	9.7	8.0

    File names   Excel: Sv34.xls
                      Minitab: Sv34.mtp
                      SPSS: Sv34.sav
                      TI-83 Plus and TI-84 Plus/ASCII: Sv34.txt

35. **Licensed Child Care Slots (Single Variable)**
The following data represent the number of licensed childcare slots in a random sample of 15 Denver neighborhoods.
Reference: The Piton Foundation, Denver, Colorado

```
523  106  184  121  357  319  656  170
241  226  741  172  266  423  212
```

File names   Excel: Sv35.xls
                    Minitab: Sv35.mtp
                    SPSS: Sv35.sav
                    TI-83 Plus and TI-84 Plus/ASCII: Sv35.txt

36. **Subsidized Housing (Single Variable)**
The following data represent the percentages of subsidized housing in a random sample of 14 Denver neighborhoods.
Reference: The Piton Foundation, Denver, Colorado

```
10.2  11.8   9.7  22.3   6.8  10.4  11.0
 5.4   6.6  13.7  13.6   6.5  16.0  24.8
```

File names   Excel: Sv36.xls
                    Minitab: Sv36.mtp
                    SPSS: Sv36.sav
                    TI-83 Plus and TI-84 Plus/ASCII: Sv36.txt

37. **Sulfate in Ground Water (Single Variable)**
The following data represent naturally occurring amounts of sulfate (SO_4) in well water. Units: parts per million. The data are from a random sample of 24 water wells in Northwest Texas.
Reference: Nichols, C.E. and Kane, V.E., Union Carbide Corporation Technical Report K/UR-1

```
1850  1150  1340  1325  2500  1060  1220  2325  460
2000  1500  1775   620  1950   780   840  2650  975
 860   495  1900  1220  2125   990
```

File names   Excel: Sv37.xls
                    Minitab: Sv37.mtp
                    SPSS: Sv37.sav
                    TI-83 Plus and TI-84 Plus/ASCII: Sv37.txt

38. **Earth's Rotation Rate (Single Variable)**
The following data represent changes in the earth's rotation (i.e., day length). Units: 0.00001 second. The data are from a random sample of 23 years.
Reference: *Acta Astron. Sinica*, Vol. 15, pp. 79-85

```
-12  110   78  126  -35  104  111   22  -31   92
 51   36  231  -13   65  119   21  104  112  -15
137  139  101
```

File names   Excel: Sv38.xls
                    Minitab: Sv38.mtp
                    SPSS: Sv38.sav
                    TI-83 Plus and TI-84 Plus/ASCII: Sv38.txt

39. Blood Glucose (Single Variable)
The following data represent glucose levels (mg/100ml) in the blood for a random sample of 27 non-obese adult subjects.
Reference: *Diabetologia*, Vol. 16, pp. 17-24

```
 80   85   75   90   70   97   91    85   90   85
105   86   78   92   93   90   80   102   90   90
 99   93   91   86   98   86   92
```

File names   Excel: Sv39.xls
                  Minitab: Sv39.mtp
                  SPSS: Sv39.sav
                  TI-83 Plus and TI-84 Plus/ASCII: Sv39.txt

40. Plant Species (Single Variable)
The following data represent the observed number of native plant species from random samples of study plots on different islands in the Galapagos Island chain.
Reference: *Science*, Vol. 179, pp. 893-895

```
23   26   33   73   21   35   30   16    3   17
 9    8    9   19   65   12   11   89   81    7
23   95    4   37   28
```

File names   Excel: Sv40.xls
                  Minitab: Sv40.mtp
                  SPSS: Sv40.sav
                  TI-83 Plus and TI-84 Plus/ASCII: Sv40.txt

41. Apples (Single Variable)
The following data represent mean fruit weights (grams) of apples per tree for a random sample of 28 trees in an agricultural experiment.
Reference: *Aust. J. Agric Res.*, Vol. 25, p783-790

```
 85.3   86.9    96.8   108.5   113.8    87.7   94.5    99.9    92.9
 67.3   90.6   129.8    48.9   117.5   100.8   94.5    94.4    98.9
 96.0   99.4    79.1   108.5    84.6   117.5   70.0   104.4   127.1
135.0
```

File names   Excel: Sv41.xls
                  Minitab: Sv41.mtp
                  SPSS: Sv41.sav
                  TI-83 Plus and TI-84 Plus/ASCII: Sv41.txt

TWO VARIABLE INDEPENDENT SAMPLES

File name prefix: Tvis followed by the number of the data file

01. Heights of Football Players Versus Heights of Basketball Players
   (Two variable independent large samples)
   The following data represent heights in feet of 45 randomly selected pro football players and 40 randomly selected pro basketball players.
   Reference: *Sports Encyclopedia of Pro Football* and *Official NBA Basketball Encyclopedia*

$X1$ = heights (ft.) of pro football players

6.33	6.50	6.50	6.25	6.50	6.33	6.25	6.17	6.42	6.33
6.42	6.58	6.08	6.58	6.50	6.42	6.25	6.67	5.91	6.00
5.83	6.00	5.83	5.08	6.75	5.83	6.17	5.75	6.00	5.75
6.50	5.83	5.91	5.67	6.00	6.08	6.17	6.58	6.50	6.25
6.33	5.25	6.67	6.50	5.83					

$X2$ = heights (ft.) of pro basketball players

6.08	6.58	6.25	6.58	6.25	5.92	7.00	6.41	6.75	6.25
6.00	6.92	6.83	6.58	6.41	6.67	6.67	5.75	6.25	6.25
6.50	6.00	6.92	6.25	6.42	6.58	6.58	6.08	6.75	6.50
6.83	6.08	6.92	6.00	6.33	6.50	6.58	6.83	6.50	6.58

File names   Excel: Tvis01.xls
                   Minitab: Tvis01.mtp
                   SPSS: Tvis01.sav
                   TI-83 Plus and TI-84 Plus/ASCII:
                             X1 data is stored in Tvis01L1.txt
                             X2 data is stored in Tvis01L2.txt

02. Petal Length for *Iris Virginica* Versus Petal Length for *Iris Setosa*
   (Two variable independent large samples)
   The following data represent petal lengths (cm) for a random sample of 35 *iris virginica* and a random sample of 38 *iris setosa*.
   Reference: Anderson, E., *Bull. Amer. Iris Soc.*

$X1$ = petal length (cm.) *iris virginica*

5.1	5.8	6.3	6.1	5.1	5.5	5.3	5.5	6.9	5.0	4.9	6.0	4.8	6.1	5.6	5.1
5.6	4.8	5.4	5.1	5.1	5.9	5.2	5.7	5.4	4.5	6.1	5.3	5.5	6.7	5.7	4.9
4.8	5.8	5.1													

$X2$ = petal length (cm.) *iris setosa*

1.5	1.7	1.4	1.5	1.5	1.6	1.4	1.1	1.2	1.4	1.7	1.0	1.7	1.9	1.6	1.4
1.5	1.4	1.2	1.3	1.5	1.3	1.6	1.9	1.4	1.6	1.5	1.4	1.6	1.2	1.9	1.5
1.6	1.4	1.3	1.7	1.5	1.7										

File names   Excel: Tvis02.xls
                   Minitab: Tvis02.mtp
                   SPSS: Tvis02.sav
                   TI-83 Plus and TI-84 Plus/ASCII:
                             X1 data is stored in Tvis02L1.txt
                             X2 data is stored in Tvis02L2.txt

03. Sepal Width Of *Iris Versicolor* Versus *Iris Virginica*
(Two variable independent large samples)
The following data represent sepal widths (cm) for a random sample of 40 *iris versicolor* and a random sample of 42 *iris virginica*.
Reference: Anderson, E., *Bull. Amer. Iris Soc.*

$X1$ = sepal width (cm.) *iris versicolor*
3.2 3.2 3.1 2.3 2.8 2.8 3.3 2.4 2.9 2.7 2.0 3.0 2.2 2.9 2.9 3.1
3.0 2.7 2.2 2.5 3.2 2.8 2.5 2.8 2.9 3.0 2.8 3.0 2.9 2.6 2.4 2.4
2.7 2.7 3.0 3.4 3.1 2.3 3.0 2.5

$X2$ = sepal width (cm.) *iris virginica*
3.3 2.7 3.0 2.9 3.0 3.0 2.5 2.9 2.5 3.6 3.2 2.7 3.0 2.5 2.8 3.2
3.0 3.8 2.6 2.2 3.2 2.8 2.8 2.7 3.3 3.2 2.8 3.0 2.8 3.0 2.8 3.8
2.8 2.8 2.6 3.0 3.4 3.1 3.0 3.1 3.1 3.1

File names	Excel: Tvis03.xls
	Minitab: Tvis03.mtp
	SPSS: Tvis03.sav
	TI-83 Plus and TI-84 Plus/ASCII:
	X1 data is stored in Tvis03L1.txt
	X2 data is stored in Tvis03L2.txt

04. Archaeology, Ceramics (Two variable independent large samples)
The following data represent independent random samples of shard counts of painted ceramics found at the Wind Mountain archaeological site.
Reference: Woosley, A. and McIntyre, A., *Mimbres Mogollon Archaeology*, University of New Mexico Press

$X1$ = count Mogollon red on brown
52 10   8 71   7 31 24 20 17   5
16 75 25 17 14 33 13 17 12 19
67 13 35 14   3   7   9 19 16 22
 7 10   9 49   6 13 24 45 14 20
 3   6 30 41 26 32 14 33   1 48
44 14 16 15 13   8 61 11 12 16
20 39

$X2$ = count Mimbres black on white
61 21 78   9 14 12 34 54 10 15
43   9   7 67 18 18 24 54   8 10
16   6 17 14 25 22 25 13 23 12
36 10 56 35 79 69 41 36 18 25
27 27 11 13

File names	Excel: Tvis04.xls
	Minitab: Tvis04.mtp
	SPSS: Tvis04.sav
	TI-83 Plus and TI-84 Plus/ASCII:
	X1 data is stored in Tvis04L1.txt
	X2 data is stored in Tvis04L2.txt

05. Agriculture, Water Content of Soil (Two variable independent large samples)
The following data represent soil water content (% water by volume) for independent random samples of soil from two experimental fields growing bell peppers.
Reference: *Journal of Agricultural, Biological, and Environmental Statistics*, Vol. 2, No. 2, pp. 149-155

$X1$ = soil water content from field I

15.1	11.2	10.3	10.8	16.6	8.3	9.1	12.3	9.1	14.3
10.7	16.1	10.2	15.2	8.9	9.5	9.6	11.3	14.0	11.3
15.6	11.2	13.8	9.0	8.4	8.2	12.0	13.9	11.6	16.0
9.6	11.4	8.4	8.0	14.1	10.9	13.2	13.8	14.6	10.2
11.5	13.1	14.7	12.5	10.2	11.8	11.0	12.7	10.3	10.8
11.0	12.6	10.8	9.6	11.5	10.6	11.7	10.1	9.7	9.7
11.2	9.8	10.3	11.9	9.7	11.3	10.4	12.0	11.0	10.7
8.8	11.1								

$X2$ = soil water content from field II

12.1	10.2	13.6	8.1	13.5	7.8	11.8	7.7	8.1	9.2
14.1	8.9	13.9	7.5	12.6	7.3	14.9	12.2	7.6	8.9
13.9	8.4	13.4	7.1	12.4	7.6	9.9	26.0	7.3	7.4
14.3	8.4	13.2	7.3	11.3	7.5	9.7	12.3	6.9	7.6
13.8	7.5	13.3	8.0	11.3	6.8	7.4	11.7	11.8	7.7
12.6	7.7	13.2	13.9	10.4	12.8	7.6	10.7	10.7	10.9
12.5	11.3	10.7	13.2	8.9	12.9	7.7	9.7	9.7	11.4
11.9	13.4	9.2	13.4	8.8	11.9	7.1	8.5	14.0	14.2

File names   Excel: Tvis05.xls
Minitab: Tvis05.mtp
SPSS: Tvis05.sav
TI-83 Plus and TI-84 Plus/ASCII:
       X1 data is stored in Tvis05L1.txt
       X2 data is stored in Tvis05L2.txt

06. Rabies (Two variable independent small samples)
The following data represent the number of cases of red fox rabies for a random sample of 16 areas in each of two different regions of southern Germany.
Reference: Sayers, B., *Medical Informatics*, Vol. 2, pp. 11-34

$X1$ = number cases in region 1
10 2 2 5 3 4 3 3 4 0 2 6 4 8 7 4

$X2$ = number cases in region 2
1 1 2 1 3 9 2 2 4 5 4 2 2 0 0 2

File names Excel: Tvis06.xls
Minitab: Tvis06.mtp
SPSS: Tvis06.sav
TI-83 Plus and TI-84 Plus/ASCII:
       X1 data is stored in Tvis06L1.txt
       X2 data is stored in Tvis06L2.txt

07. **Weight of Football Players Versus Weight of Basketball Players**
    (Two variable independent small samples)
    The following data represent weights in pounds of 21 randomly selected pro football players, and 19 randomly selected pro basketball players.
    Reference: *Sports Encyclopedia of Pro Football* and *Official NBA Basketball Encyclopedia*

    $X1$ = weights (lb.) of pro football players
245	262	255	251	244	276	240	265	257	252	282
256	250	264	270	275	245	275	253	265	270	

    $X2$ = weights (lb.) of pro basketball players
205	200	220	210	191	215	221	216	228	207
225	208	195	191	207	196	181	193	201	

    File names   Excel: Tvis07.xls
                      Minitab: Tvis07.mtp
                      SPSS: Tvis07.sav
                      TI-83 Plus and TI-84 Plus/ASCII:
                                X1 data is stored in Tvis07L1.txt
                                X2 data is stored in Tvis07L2.txt

08. **Birth Rate (Two variable independent small samples)**
    The following data represent birth rates (per 1000 residential population) for independent random samples of counties in California and Maine.
    Reference: *County and City Data Book,* 12th edition, U.S. Dept. of Commerce

    $X1$ = birth rate in California counties
14.1	18.7	20.4	20.7	16.0	12.5	12.9	9.6	17.6
18.1	14.1	16.6	15.1	18.5	23.6	19.9	19.6	14.9
17.7	17.8	19.1	22.1	15.6				

    $X2$ = birth rate in Maine counties
15.1	14.0	13.3	13.8	13.5	14.2	14.7	11.8	13.5	13.8
16.5	13.8	13.2	12.5	14.8	14.1	13.6	13.9	15.8	

    File names   Excel: Tvis08.xls
                      Minitab: Tvis08.mtp
                      SPSS: Tvis08.sav
                      TI-83 Plus and TI-84 Plus/ASCII:
                                X1 data is stored in Tvis08L1.txt
                                X2 data is stored in Tvis08L2.txt

09. Death Rate (Two variable independent small samples)
The following data represent death rates (per 1000 resident population) for independent random samples of counties in Alaska and Texas.
Reference: *County and City Data Book,* 12th edition, U.S. Dept. of Commerce

X1 = death rate in Alaska counties
 1.4   4.2   7.3   4.8   3.2   3.4   5.1   5.4
 6.7   3.3   1.9   8.3   3.1   6.0   4.5   2.5

X2 = death rate in Texas counties
 7.2   5.8   10.5   6.6   6.9   9.5   8.6   5.9   9.1
 5.4   8.8   6.1   9.5   9.6   7.8   10.2   5.6   8.6

File names   Excel: Tvis09.xls
                       Minitab: Tvis09.mtp
                       SPSS: Tvis09.sav
                       TI-83 Plus and TI-84 Plus/ASCII:
                                X1 data is stored in Tvis09L1.txt
                                X2 data is stored in Tvis09L2.txt

10. Pickup Trucks (Two variable independent small samples)
The following data represent the retail prices (in thousands of dollars) for independent random samples of models of pickup trucks.
Reference: *Consumer Guide,* Vol. 681

X1 = prices for different GMC Sierra 1500 models
 17.4   23.3   29.2   19.2   17.6   19.2   23.6   19.5   22.2
 24.0   26.4   23.7   29.4   23.7   26.7   24.0   24.9

X2 = prices for different Chevrolet Silverado 1500 models
 17.5   23.7   20.8   22.5   24.3   26.7   24.5   17.8
 29.4   29.7   20.1   21.1   22.1   24.2   27.4   28.1

File names   Excel: Tvis10.xls
                       Minitab: Tvis10.mtp
                       SPSS: Tvis10.sav
                       TI-83 Plus and TI-84 Plus/ASCII:
                                X1 data is stored in Tvis10L1.txt
                                X2 data is stored in Tvis10L2.txt

TWO VARIABLE DEPENDENT SAMPLES

File name prefix: Tvds followed by the number of the data file

01. Average Faculty Salary, Male versus Female (Two variable dependent samples)
In the following data pairs, A = average salaries for males ($1000/yr) and B = average salaries for females ($1000/yr) for assistant professors at the same college or university. A random sample of 22 U.S. colleges and universities was used.
Reference: *Academe, Bulletin of the American Association of University Professors*

A: 34.5 30.5 35.1 35.7 31.5 34.4 32.1 30.7 33.7 35.3
B: 33.9 31.2 35.0 34.2 32.4 34.1 32.7 29.9 31.2 35.5

A: 30.7 34.2 39.6 30.5 33.8 31.7 32.8 38.5 40.5 25.3
B: 30.2 34.8 38.7 30.0 33.8 32.4 31.7 38.9 41.2 25.5

A: 28.6 35.8
B: 28.0 35.1

File names   Excel: Tvds01.xls
                    Minitab: Tvds01.mtp
                    SPSS: Tvds01.sav
                    TI-83 Plus and TI-84 Plus/ASCII:
                              X1 data is stored in Tvds01L1.txt
                              X2 data is stored in Tvds01L2.txt

**02. Unemployment for College Graduates Versus High School Only
(Two variable dependent samples)**
In the following data pairs, A = percent unemployment for college graduates and B = percent unemployment for high school only graduates. The data are paired by year.
Reference: *Statistical Abstract of the United States*

A: 2.8 2.2 2.2 1.7 2.3 2.3 2.4 2.7 3.5 3.0 1.9 2.5
B: 5.9 4.9 4.8 5.4 6.3 6.9 6.9 7.2 10.0 8.5 5.1 6.9

File names   Excel: Tvds02.xls
                    Minitab: Tvds02.mtp
                    SPSS: Tvds02.sav
                    TI-83 Plus and TI-84 Plus/ASCII:
                              X1 data is stored in Tvds02L1.txt
                              X2 data is stored in Tvds02L2.txt

03. **Number of Navajo Hogans versus Modern Houses (Two variable dependent samples)**
In the following data pairs, A = number of traditional Navajo hogans in a given district and B = number of modern houses in a given district. The data are paired by district of the Navajo reservation. A random sample of 8 districts was used.
Reference: *Navajo Architecture, Forms, History, Distributions* by S.C. Jett and V.E. Spencer, Univ. of Arizona Press

A:	13	14	46	32	15	47	17	18
B:	18	16	68	9	11	28	50	50

File names   Excel: Tvds03.xls
                    Minitab: Tvds03.mtp
                    SPSS: Tvds03.sav
                    TI-83 Plus and TI-84 Plus/ASCII:
                              X1 data is stored in Tvds03L1.txt
                              X2 data is stored in Tvds03L2.txt

04. **Temperatures in Miami versus Honolulu (Two variable dependent samples)**
In the following data pairs, A = average monthly temperature in Miami and B = average monthly temperature in Honolulu. The data are paired by month.
Reference: U.S. Department of Commerce Environmental Data Service

A:	67.5	68.0	71.3	74.9	78.0	80.9	82.2	82.7	81.6	77.8	72.3	68.5
B:	74.4	72.6	73.3	74.7	76.2	78.0	79.1	79.8	79.5	78.4	76.1	73.7

File names   Excel: Tvds04.xls
                    Minitab: Tvds04.mtp
                    SPSS: Tvds04.sav
                    TI-83 Plus and TI-84 Plus/ASCII:
                              X1 data is stored in Tvds04L1.txt
                              X2 data is stored in Tvds04L2.txt

05. **January/February Ozone Column (Two variable dependent samples)**
In the following pairs, the data represent the thickness of the ozone column in Dobson units: one milli-centimeter of ozone at standard temperature and pressure.
    A = monthly mean thickness in January
    B = monthly mean thickness in February
The data are paired by year for a random sample of 15 years.
Reference: Laboratorium für Atmosphärensphysik, Switzerland

A:	360	324	377	336	383	361	369	349
B:	365	325	359	352	397	351	367	397

A:	301	354	344	329	337	387	378
B:	335	338	349	393	370	400	411

File names   Excel: Tvds05.xls
                    Minitab: Tvds05.mtp
                    SPSS: Tvds05.sav
                    TI-83 Plus and TI-84 Plus/ASCII:
                              X1 data is stored in Tvds05L1.txt
                              X2 data is stored in Tvds05L2.txt

06. Birth Rate/Death Rate (Two variable dependent samples)
In the following data pairs, A = birth rate (per 1000 resident population) and B = death rate (per 1000 resident population). The data are paired by county in Iowa.
Reference: *County and City Data Book*, 12th edition, U.S. Dept. of Commerce

A:	12.7	13.4	12.8	12.1	11.6	11.1	14.2
B:	9.8	14.5	10.7	14.2	13.0	12.9	10.9

A:	12.5	12.3	13.1	15.8	10.3	12.7	11.1
B:	14.1	13.6	9.1	10.2	17.9	11.8	7.0

File names   Excel: Tvds06.xls
                        Minitab: Tvds06.mtp
                        SPSS: Tvds06.sav
                        TI-83 Plus and TI-84 Plus/ASCII:
                                    X1 data is stored in Tvds06L1.txt
                                    X2 data is stored in Tvds06L2.txt

07. Democrat/Republican (Two variable dependent samples)
In the following data pairs A = percentage of voters who voted Democrat and B = percentage of voters who voted Republican in a recent national election. The data are paired by county in Indiana.
Reference: *County and City Data Book*, 12th edition, U.S. Dept. of Commerce

A:	42.2	34.5	44.0	34.1	41.8	40.7	36.4	43.3	39.5
B:	35.4	45.8	39.4	40.0	39.2	40.2	44.7	37.3	40.8

A:	35.4	44.1	41.0	42.8	40.8	36.4	40.6	37.4
B:	39.3	36.8	35.5	33.2	38.3	47.7	41.1	38.5

File names   Excel: Tvds07.xls
                        Minitab: Tvds07.mtp
                        SPSS: Tvds07.sav
                        TI-83 Plus and TI-84 Plus/ASCII:
                                    X1 data is stored in Tvds07L1.txt
                                    X2 data is stored in Tvds07L2.txt

08. Santiago Pueblo Pottery (Two variable dependent samples)
In the following data, A = percentage of utility pottery and B = percentage of ceremonial pottery found at the Santiago Pueblo archaeological site. The data are paired by location of discovery.
Reference: Laboratory of Anthropology, Notes 475, Santa Fe, New Mexico

A:	41.4	49.6	55.6	49.5	43.0	54.6	46.8	51.1	43.2	41.4
B:	58.6	50.4	44.4	59.5	57.0	45.4	53.2	48.9	56.8	58.6

File names   Excel: Tvds08.xls
                        Minitab: Tvds08.mtp
                        SPSS: Tvds08.sav
                        TI-83 Plus and TI-84 Plus/ASCII:
                                    X1 data is stored in Tvds08L1.txt
                                    X2 data is stored in Tvds08L2.txt

09. Poverty Level (Two variable dependent samples)
In the following data pairs, A = percentage of population below poverty level in 1998 and B = percentage of population below poverty level in 1990. The data are grouped by state and District of Columbia.
Reference: *Statistical Abstract of the United States*, 120th edition

A:	14.5	9.4	16.6	14.8	15.4	9.2	9.5	10.3	22.3	13.1
B:	19.2	11.4	13.7	19.6	13.9	13.7	6.0	6.9	21.1	14.4

A:	13.6	10.9	13.0	10.1	9.4	9.1	9.6	13.5	19.1	10.4
B:	15.8	11.0	14.9	13.7	13.0	10.4	10.3	17.3	23.6	13.1

A:	7.2	8.7	11.0	10.4	17.6	9.8	16.6	12.3	10.6	9.8
B:	9.9	10.7	14.3	12.0	25.7	13.4	16.3	10.3	9.8	6.3

A:	8.6	20.4	16.7	14.0	15.1	11.2	14.1	15.0	11.2	11.6
B:	9.2	20.9	14.3	13.0	13.7	11.5	15.6	9.2	11.0	7.5

A:	13.7	10.8	13.4	15.1	9.0	9.9	8.8	8.9	17.8	8.8	10.6
B:	16.2	13.3	16.9	15.9	8.2	10.9	11.1	8.9	18.1	9.3	11.0

File names   Excel: Tvds09.xls
Minitab: Tvds09.mtp
SPSS: Tvds09.sav
TI-83 Plus and TI-84 Plus/ASCII:
X1 data is stored in Tvds09L1.txt
X2 data is stored in Tvds09L2.txt

10. Cost of Living Index (Two variable dependent samples)
The following data pairs represent cost of living index for A = grocery items and B = health care. The data are grouped by metropolitan areas.
Reference: *Statistical Abstract of the United States*, 120th edition

A:	96.6	97.5	113.9	88.9	108.3	99.0	97.3	87.5	96.8
B:	91.6	95.9	114.5	93.6	112.7	93.6	99.2	93.2	105.9

A:	102.1	114.5	100.9	100.0	100.7	99.4	117.1	111.3	102.2
B:	110.8	127.0	91.5	100.5	104.9	104.8	124.1	124.6	109.1

A:	95.3	91.1	95.7	87.5	91.8	97.9	97.4	102.1	94.0
B:	98.7	95.8	99.7	93.2	100.7	96.0	99.6	98.4	94.0

A:	115.7	118.3	101.9	88.9	100.7	99.8	101.3	104.8	100.9
B:	121.2	122.4	110.8	81.2	104.8	109.9	103.5	113.6	94.6

A:	102.7	98.1	105.3	97.2	105.2	108.1	110.5	99.3	99.7
B:	109.8	97.6	109.8	107.4	97.7	124.2	110.9	106.8	94.8

File names   Excel: Tvds10.xls
Minitab: Tvds10.mtp
SPSS: Tvds10.sav
TI-83 Plus and TI-84 Plus/ASCII:
X1 data is stored in Tvds10L1.txt
X2 data is stored in Tvds10L2.txt

SIMPLE LINEAR REGRESSION

File name prefix: Slr followed by the number of the data file

01. List Price versus Best Price for a New GMC Pickup Truck (Simple Linear Regression)
In the following data pairs, X = list price (in $1000) for a GMC pickup truck and Y = best price (in $1000) for a GMC pickup truck.
Reference: *Consumer's Digest*

X: 12.4 14.3 14.5 14.9 16.1 16.9 16.5 15.4 17.0 17.9
Y: 11.2 12.5 12.7 13.1 14.1 14.8 14.4 13.4 14.9 15.6

X: 18.8 20.3 22.4 19.4 15.5 16.7 17.3 18.4 19.2 17.4
Y: 16.4 17.7 19.6 16.9 14.0 14.6 15.1 16.1 16.8 15.2

X: 19.5 19.7 21.2
Y: 17.0 17.2 18.6

File names   Excel: Slr01.xls
                    Minitab: Slr01.mtp
                    SPSS: Slr01.sav
                    TI-83 Plus and TI-84 Plus/ASCII:
                            X1 data is stored in Slr01L1.txt
                            X2 data is stored in Slr01L2.txt

02. Cricket Chirps versus Temperature (Simple Linear Regression)
In the following data pairs, X = chirps/sec for the striped ground cricket and Y = temperature in degrees Fahrenheit.
Reference: *The Song of Insects* by Dr. G.W. Pierce, Harvard College Press

X: 20.0 16.0 19.8 18.4 17.1 15.5 14.7 17.1
Y: 88.6 71.6 93.3 84.3 80.6 75.2 69.7 82.0

X: 15.4 16.2 15.0 17.2 16.0 17.0 14.4
Y: 69.4 83.3 79.6 82.6 80.6 83.5 76.3

File names   Excel: Slr02.xls
                    Minitab: Slr02.mtp
                    SPSS: Slr02.sav
                    TI-83 Plus and TI-84 Plus/ASCII:
                            X1 data is stored in Slr02L1.txt
                            X2 data is stored in Slr02L2.txt

03. Diameter of Sand Granules versus Slope on Beach (Simple Linear Regression)
In the following data pairs, X = median diameter (mm) of granules of sand and Y = gradient of beach slope in degrees. The data are for naturally occurring ocean beaches.
Reference: *Physical Geography* by A.M. King, Oxford Press, England

X:	0.170	0.190	0.220	0.235	0.235	0.300	0.350	0.420	0.850
Y:	0.630	0.700	0.820	0.880	1.150	1.500	4.400	7.300	11.300

File names

Excel: Slr03.xls
Minitab: Slr03.mtp
SPSS: Slr03.sav
TI-83 Plus and TI-84 Plus/ASCII:
    X1 data is stored in Slr03L1.txt
    X2 data is stored in Slr03L2.txt

04. National Unemployment Male versus Female (Simple Linear Regression)
In the following data pairs, X = national unemployment rate for adult males and Y = national unemployment rate for adult females.
Reference: *Statistical Abstract of the United States*

X:	2.9	6.7	4.9	7.9	9.8	6.9	6.1	6.2	6.0	5.1	4.7	4.4	5.8
Y:	4.0	7.4	5.0	7.2	7.9	6.1	6.0	5.8	5.2	4.2	4.0	4.4	5.2

File names

Excel: Slr04.xls
Minitab: Slr04.mtp
SPSS: Slr04.sav
TI-83 Plus and TI-84 Plus/ASCII:
    X1 data is stored in Slr04L1.txt
    X2 data is stored in Slr04L2.txt

05. Fire and Theft in Chicago (Simple Linear Regression)
In the following data pairs, X = fires per 1000 housing units and Y = thefts per 1000 population within the same zip code in the Chicago metro area.
Reference: U.S. Commission on Civil Rights

X:	6.2	9.5	10.5	7.7	8.6	34.1	11.0	6.9	7.3	15.1
Y:	29	44	36	37	53	68	75	18	31	25

X:	29.1	2.2	5.7	2.0	2.5	4.0	5.4	2.2	7.2	15.1
Y:	34	14	11	11	22	16	27	9	29	30

X:	16.5	18.4	36.2	39.7	18.5	23.3	12.2	5.6	21.8	21.6
Y:	40	32	41	147	22	29	46	23	4	31

X:	9.0	3.6	5.0	28.6	17.4	11.3	3.4	11.9	10.5	10.7
Y:	39	15	32	27	32	34	17	46	42	43

X:	10.8	4.8
Y:	34	19

File names

Excel: Slr05.xls
Minitab: Slr05.mtp
SPSS: Slr05.sav
TI-83 Plus and TI-84 Plus/ASCII:
    X1 data is stored in Slr05L1.txt
    X2 data is stored in Slr05L2.txt

06.   Auto Insurance in Sweden (Simple Linear Regression)
In the following data pairs, X = number of claims and Y = total payment for all the claims in thousands of Swedish Kronor for geographical zones in Sweden.
Reference: Swedish Committee on Analysis of Risk Premium in Motor Insurance

X:	108	19	13	124	40	57	23	14	45	10
Y:	392.5	46.2	15.7	422.2	119.4	170.9	56.9	77.5	214.0	65.3

X:	5	48	11	23	7	2	24	6	3	23
Y:	20.9	248.1	23.5	39.6	48.8	6.6	134.9	50.9	4.4	113.0

X:	6	9	9	3	29	7	4	20	7	4
Y:	14.8	48.7	52.1	13.2	103.9	77.5	11.8	98.1	27.9	38.1

X:	0	25	6	5	22	11	61	12	4	16
Y:	0.0	69.2	14.6	40.3	161.5	57.2	217.6	58.1	12.6	59.6

X:	13	60	41	37	55	41	11	27	8	3
Y:	89.9	202.4	181.3	152.8	162.8	73.4	21.3	92.6	76.1	39.9

X:	17	13	13	15	8	29	30	24	9	31
Y:	142.1	93.0	31.9	32.1	55.6	133.3	194.5	137.9	87.4	209.8

X:	14	53	26
Y:	95.5	244.6	187.5

File names   Excel: Slr06.xls
                 Minitab: Slr06.mtp
                 SPSS: Slr06.sav
                 TI-83 Plus and TI-84 Plus/ASCII:
                          X1 data is stored in Slr06L1.txt
                          X2 data is stored in Slr06L2.txt

07.   Gray Kangaroos (Simple Linear Regression)
In the following data pairs, X = nasal length (mm × 10) and Y = nasal width (mm × 10) for a male gray kangaroo from a random sample of such animals.
Reference: *Australian Journal of Zoology*, Vol. 28, pp. 607-613

X:	609	629	620	564	645	493	606	660	630	672
Y:	241	222	233	207	247	189	226	240	215	231

X:	778	616	727	810	778	823	755	710	701	803
Y:	263	220	271	284	279	272	268	278	238	255

X:	855	838	830	864	635	565	562	580	596	597
Y:	308	281	288	306	236	204	216	225	220	219

X:	636	559	615	740	677	675	629	692	710	730
Y:	201	213	228	234	237	217	211	238	221	281

X:	763	686	717	737	816
Y:	292	251	231	275	275

File names   Excel: Slr07.xls
Minitab: Slr07.mtp
SPSS: Slr07.sav
TI-83 Plus and TI-84 Plus/ASCII:
        X1 data is stored in Slr07L1.txt
        X2 data is stored in Slr07L2.txt

08. Pressure and Weight in Cryogenic Flow Meters (Simple Linear Regression)

In the following data pairs, X = pressure (lb./sq. in.) of liquid nitrogen and Y = weight in pounds of liquid nitrogen passing through a flow meter each second.
Reference: *Technometrics*, Vol. 19, pp. 353-379

X:	75.1	74.3	88.7	114.6	98.5	112.0	114.8	62.2	107.0
Y:	577.8	577.0	570.9	578.6	572.4	411.2	531.7	563.9	406.7

X:	90.5	73.8	115.8	99.4	93.0	73.9	65.7	66.2	77.9
Y:	507.1	496.4	505.2	506.4	510.2	503.9	506.2	506.3	510.2

X:	109.8	105.4	88.6	89.6	73.8	101.3	120.0	75.9	76.2
Y:	508.6	510.9	505.4	512.8	502.8	493.0	510.8	512.8	513.4

X:	81.9	84.3	98.0
Y:	510.0	504.3	522.0

File names   Excel: Slr08.xls
Minitab: Slr08.mtp
SPSS: Slr08.sav
TI-83 Plus and TI-84 Plus/ASCII:
        X1 data is stored in Slr08L1.txt
        X2 data is stored in Slr08L2.txt

09. Ground Water Survey (Simple Linear Regression)

In the following data pairs, X = pH of well water and Y = bicarbonate (parts per million) of well water. The data are for well water from a random sample of wells in Northwest Texas.
Reference: Nichols, C.E. and Kane, V.E., Union Carbide Technical Report K/UR-1

X:	7.6	7.1	8.2	7.5	7.4	7.8	7.3	8.0	7.1	7.5
Y:	157	174	175	188	171	143	217	190	142	190

X:	8.1	7.0	7.3	7.8	7.3	8.0	8.5	7.1	8.2	7.9
Y:	215	199	262	105	121	81	82	210	202	155

X:	7.6	8.8	7.2	7.9	8.1	7.7	8.4	7.4	7.3	8.5
Y:	157	147	133	53	56	113	35	125	76	48

X:	7.8	6.7	7.1	7.3
Y:	147	117	182	87

File names   Excel: Slr09.xls
Minitab: Slr09.mtp
SPSS: Slr09.sav
TI-83 Plus and TI-84 Plus/ASCII:
        X1 data is stored in Slr09L1.txt
        X2 data is stored in Slr09L2.txt

10.   *Iris Setosa* **(Simple Linear Regression)**
    In the following data pairs, X = sepal width (cm) and Y = sepal length (cm). The data are for a random sample of the wild flower *iris setosa*.
    Reference: Fisher, R.A., *Ann. Eugenics,* Vol. 7, Part II, pp. 179-188

X:	3.5	3.0	3.2	3.1	3.6	3.9	3.4	3.4	2.9	3.1
Y:	5.1	4.9	4.7	4.6	5.0	5.4	4.6	5.0	4.4	4.9

X:	3.7	3.4	3.0	4.0	4.4	3.9	3.5	3.8	3.8	3.4
Y:	5.4	4.8	4.3	5.8	5.7	5.4	5.1	5.7	5.1	5.4

X:	3.7	3.6	3.3	3.4	3.0	3.4	3.5	3.4	3.2	3.1
Y:	5.1	4.6	5.1	4.8	5.0	5.0	5.2	5.2	4.7	4.8

X:	3.4	4.1	4.2	3.1	3.2	3.5	3.6	3.0	3.4	3.5
Y:	5.4	5.2	5.5	4.9	5.0	5.5	4.9	4.4	5.1	5.0

X:	2.3	3.2	3.5	3.8	3.0	3.8	3.7	3.3
Y:	4.5	4.4	5.0	5.1	4.8	4.6	5.3	5.0

File names	Excel: Slr10.xls
	Minitab: Slr10.mtp
	SPSS: Slr10.sav
	TI-83 Plus and TI-84 Plus/ASCII:
	X1 data is stored in Slr10L1.txt
	X2 data is stored in Slr10L2.txt

11.   **Pizza Franchise (Simple Linear Regression)**
    In the following data pairs, X = annual franchise fee ($1000) and Y = start-up cost ($1000) for a pizza franchise.
    Reference: *Business Opportunities Handbook*

X:	25.0	8.5	35.0	15.0	10.0	30.0	10.0	50.0	17.5	16.0
Y:	125	80	330	58	110	338	30	175	120	135

X:	18.5	7.0	8.0	15.0	5.0	15.0	12.0	15.0	28.0	20.0
Y:	97	50	55	40	35	45	75	33	55	90

X:	20.0	15.0	20.0	25.0	20.0	3.5	35.0	25.0	8.5	10.0
Y:	85	125	150	120	95	30	400	148	135	45

X:	10.0	25.0
Y:	87	150

File names	Excel: Slr11.xls
	Minitab: Slr11.mtp
	SPSS: Slr11.sav
	TI-83 Plus and TI-84 Plus/ASCII:
	X1 data is stored in Slr11L1.txt
	X2 data is stored in Slr11L2.txt

12. **Prehistoric Pueblos (Simple Linear Regression)**
In the following data pairs, X = estimated year of initial occupation and Y = estimated year of end of
occupation. The data are for each prehistoric pueblo in a random sample of such pueblos in Utah,
Arizona, and Nevada.
Reference: *Prehistoric Pueblo World*, by A. Adler, Univ. of Arizona Press

X:	1000	1125	1087	1070	1100	1150	1250	1150	1100
Y:	1050	1150	1213	1275	1300	1300	1400	1400	1250

X:	1350	1275	1375	1175	1200	1175	1300	1260	1330
Y:	1830	1350	1450	1300	1300	1275	1375	1285	1400

X:	1325	1200	1225	1090	1075	1080	1080	1180	1225
Y:	1400	1285	1275	1135	1250	1275	1150	1250	1275

X:	1175	1250	1250	750	1125	700	900	900	850
Y:	1225	1280	1300	1250	1175	1300	1250	1300	1200

File names   Excel: Slr12.xls
                    Minitab: Slr12.mtp
                    SPSS: Slr12.sav
                    TI-83 Plus and TI-84 Plus/ASCII:
                              X1 data is stored in Slr12L1.txt
                              X2 data is stored in Slr12L2.txt

Made in the USA
Monee, IL
07 July 2026

56552159R00070